MYSTICISM AND THE NEW PHYSICS

Michael Talbot is a freelance writer whose work has appeared in publications ranging from *Omni* to *The Village Voice*. He has been exploring the confluence between science and the spiritual for over twenty years and is the author of *Beyond the Quantum*, *Your Past Lives: a Reincarnation Handbook* and three novels. He lives and writes in New York City.

MYSTICISM AND THE NEW PHYSICS

Michael Talbot

ARKANA
PENGUIN BOOKS

ARKANA

Published by the Penguin Group
Penguin Books Ltd, 80 Strand, London WC2R 0RL, England
Penguin Putnam Inc., 375 Hudson Street, New York, New York 10014, USA
Penguin Books Australia Ltd, 250 Camberwell Road, Camberwell, Victoria 3124, Australia
Penguin Books Canada Ltd, 10 Alcorn Avenue, Toronto, Ontario, Canada M4V 3B2
Penguin Books India (P) Ltd, 11 Community Centre, Panchsheel Park, New Delhi – 110 017, India
Penguin Books (NZ) Ltd, Cnr Rosedale and Airborne Roads, Albany, Auckland, New Zealand
Penguin Books (South Africa) (Pty) Ltd, 24 Sturdee Avenue, Rosebank 2196, South Africa

Penguin Books Ltd, Registered Offices: 80 Strand, London WC2R 0RL, England

www.penguin.com

First published by Routledge & Kegan Paul 1981
Revised and updated edition published by Arkana 1993

032

Printed and bound in Great Britain by Clays Ltd, Elcograf S.p.A.
Filmset in 10/12 pt Monophoto Bembo

ISBN-13: 978–0–140–19328–2

www.greenpenguin.co.uk

MIX
Paper | Supporting
responsible forestry
FSC
www.fsc.org   FSC® C018179

Penguin Books is committed to a sustainable
future for our business, our readers and our planet.
This book is made from Forest Stewardship
Council™ certified paper.

For my mother and father,
Nancy Caroline and Frederick Bernard Talbot

Gracious one play, your head is an empty shell
wherein your mind frolics infinitely.

—AN OLD SANSKRIT PROVERB

Contents

PART THREE *MYSTICISM AND THE NEW PHYSICS*

Introduction

Let us admit what all idealists admit — the hallucinatory nature of the world. Let us do what no idealist has done — let us search for unrealities that confirm that nature. I believe we shall find them in the antinomies of Kant and in the dialectic of Zeno . . . . 'The greatest wizard (Novalis writes memorably) would be the one who bewitched himself to the point of accepting his own phantasmagorias as autonomous apparitions. Wouldn't that be our case.' I surmise it is so. We (that indivisible divinity that operates in us) have dreamed the world. We have dreamed it as enduring, mysterious, visible, omnipresent in space and stable in time; but we have consented to tenuous and eternal intervals of illogicalness in its architecture that we might know it is false.

JORGE LUIS BORGES, *Other Inquisitions*

In this quote the pre-eminent Argentinian writer Borges presents a view normally held by mystics and 'idealists': that is, the hallucinatory nature of the world. We have dreamed it, Borges states simply.

In an article published in 1976 I suggested that persistent reports from various peoples throughout history of encounters with supernatural beings — folkloric personages, UFO entities, appearances of the Virgin, and so forth — forces us to reach a similar conclusion. On one hand such phenomena possess traits that suggest they are a product of the myths and beliefs of the individuals witnessing them and are some kind of psychological projection. But on the other, they also occasionally seem quite real, with UFOs that can be tracked on radar, or manifestations of the Virgin (such as those that occurred in Knock, Ireland, in 1879 and Zeitoun, Egypt, in 1968) which were witnessed by dozens and even hundreds of people, and so on.

In the article I suggested that such phenomena cannot be understood as either purely objective or subjective events. Rather they are somehow both, or 'omnijective'. Furthermore, their existence suggests that in the universe at large there is also ultimately no division between mind and reality. Despite its apparent materiality the physical world is also omnijective. I discuss this idea at length in my book *The Holographic Universe* (published in the United States by Harper Collins and in Great Britain by Grafton (1991)).

An omnijective concept of the universe is by no means new. Over two thousand years ago the Hindu Tantric tradition postulated a similar philosophy. According to Tantra, reality is illusion, or maya. The major error we commit in not perceiving this maya, say the Tantras, is that we perceive ourselves as separate from our environment. The Tantras are very explicit on this point. The observer and objective reality are one.

When we dream, the omnijective nature of the dream is obvious. I may dream that I am sitting at a table having breakfast and talking with my friends, but when I awake, I know that both I and my friends are part of the continuum of the dream. To say that there are many 'consciousnesses' in the dream is merely a semantic distinction. All the people in the dream are maya. They are constructions of consciousness.

Alfred North Whitehead postulated a similar dreamlike nature to reality: '. . . [the] theory which I am arguing against is to bifurcate nature into two divisions, namely into the nature apprehended in awareness and the nature which is the cause of awareness. The nature which is the fact apprehended in awareness holds within it the greenness of the trees, the song of the birds, the warmth of the sun, the hardness of the chairs, and the feel of the velvet. The nature which is the cause of awareness is the conjectured system of molecules and electrons which so affects the mind as to produce the awareness of apparent nature. The meeting point of these two natures is the mind . . .'(81)*

Indeed, there is a vast philosophical and metaphysical tradition behind the philosophy that the universe is omnijective. The mystics tell us this is true. The idealists tell us it is true. Most exciting of all, the physicists tell us it is true. As Jack Sarfatti states, in *Psychoenergetic Systems*, 'The full meaning of quantum theory is still in the stage of being born. In my opinion, the quantum principle involves *mind* in an essential way along the lines suggested by Parmenides, Bishop Berkeley, Jeans, Whitehead *et al.*'(67)

*For Notes see pp. 171–5

In 1927 Werner Heisenberg presented his famous Uncertainty Principle and started a philosophical debate among the quantum physicists that still has not resolved itself. In vastly simplified terms Heisenberg stated that the observer alters the observed by the mere act of observation. In striving to penetrate the secrets of matter Heisenberg perhaps unwittingly caught a glimpse of maya – one of the 'illogicalnesses' in the architecture of the universe that Borges speaks of. As Heisenberg stated, 'The conception of objective reality . . . evaporated into the . . . mathematics that represents no longer the behavior of elementary particles but rather our knowledge of this behavior.'(41)

The most astounding transformation of world view that the new physics has undertaken is this – the recognition that *consciousness does play a role in the so-called physical universe.* Since the time of Newton physics has always tried to maintain a strictly empirical approach. The *raison d'être* of the old physics was that there was a physical world available to direct touch. It was a trusted myth that the laws of the physical world did not change; given the proper tools and instruction any physicist could duplicate the experiments and observations of any other physicist. The role of empiricism in science has always demanded a dispassionate observer and concentrated upon objective reality as a single, observable 'something' *a priori* to the consciousness. It doesn't matter which physicist or which mind makes the observation. It's the 'same' universe and that's what counts.

But the new physics, the physics of quantum theory – the branch of physics which deals with very small 'quantities' of matter and energy – has found *it does matter.* Given the proper tools and instruction a physicist will not necessarily duplicate the experiments and observations of another physicist. The outcome of any particular experiment no longer seems to depend only upon the 'laws' of the physical world, but also upon the consciousness of the observer. Indeed, as Princeton physicist John A. Wheeler suggests, we must replace the term 'observer' with the term 'participator'.(79) We cannot *observe* the physical world, for as the new physics tell us, there is no one physical world. We *participate* within a spectrum of all possible realities.

The recognition of the role of consciousness in the processes of the physical universe is a radical departure from classical physics. But it is what the mystics have been telling us all along. This, then, is the guiding topic of this book – to point out the confluence of mysticism and the new physics; to offer a perspective of the universe under this new and subsuming structure; and to point out the radical, even awesome implications of such a perspective.

We should pay special attention to the words of the quantum physicists, for they are opening a Pandora's box when they admit that the human mind has a hand in the phenomena of the so-called objective world. A new beast has stuck its toes inside the door of classical physics and it will still be many years before we see the full effect this will have on the world of sticks and stones. One thing is certain: if the human mind has an effect on even so much as a single particle, the entire ecology of the material universe is affected. Our view of reality is in the first sluggish pangs of a radical change.

Over half a century has elapsed since Heisenberg formulated his revolutionary Uncertainty Principle and even today the insights of the new physics are just beginning to trickle down from the top of the information pyramid. The implications of the confluence of mysticism and physics are that all of our notions about the *absoluteness* of the physical universe are wrong. Slowly and painfully we are realizing the obvious: our concepts are based upon a most intriguing maya. Our constructs need amending. The very epistemological foundations of our environment and ourselves must shift as our prejudices are attacked. As Heisenberg stated, 'The violent reaction on the recent development of modern physics can only be understood when one realizes that here the foundations of physics have started moving; and that this motion has caused the feeling that the ground would be cut from science.'(41)

Just as the shift from Newtonian to Einsteinian physics brought about no death, the confluence to mysticism and the new physics will bring about no death of the study of physics, only transformation. For there are no ends in human study – only a continual flux and change as old systems are subsumed by larger and larger hierarchies. The confluence is very much, then, a melding, a synthesis – two globules of mercury touching to produce an even larger globule. Perhaps when the scientific establishment at large realizes that the puzzles encountered in psychic phenomena are already part of the very fabric of science, serious research efforts can begin. Indeed, in light of Wheeler's notion of the role of the 'participator', physics might have to invent psychic research, if it did not already exist.

Any work dealing with the confluence of mysticism and physics must necessarily be cautious. As Whitehead warned, 'There will be some fundamental assumptions which adherents of all the variant systems within the epoch unconsciously presuppose. Such assumptions appear so obvious that people do not know what they are assuming because no other way of putting things has ever occurred to them.'(82)

For instance, our experience tells us that the universe is Euclidean. Given a straight line and a point beside that line, our intuitions tell us that there is only one possible line that passes through the point and is parallel to the first line.

This is such a fundamental assumption that most of us find it difficult to conceptualize any other possibility. In the first quarter of the nineteenth century, a Hungarian and a Russian, Bolyai and Lobachevski, questioned the mathematical 'truth' of Euclid's postulate. They ultimately demonstrated that the postulate cannot be proven. By assuming from the beginning that through a given point *two* lines can be drawn parallel to the first line, Lobachevski constructed a geometry whose logic is as faultless as that of Euclidean geometry.

In *Foundations of Science*, the French mathematician Henri Poincaré pointed out that 'the axioms of geometry therefore are neither *a priori* judgments nor experimental facts. They are conventions; our choice among all possible conventions is guided by experimental facts; but it remains free and is limited only by the necessity of avoiding all contradictions.'

The belief that Euclidean geometry – the geometry which is still taught as gospel in school systems – is the only possible geometrical system was shown to be unfounded. In the nineteenth century the mathematician Georg Riemann constructed a system based upon the proposition that there are *no* lines that pass through the point and are parallel to the first line. Although Riemann introduced this third geometry as a purely abstract mathematical idea, Einstein ultimately resorted to Riemannian geometry for the mathematical framework of his theory of relativity. On the astronomical level, in phenomena such as gravitational collapse and blackholes, Euclidean geometry simply doesn't work. Our prejudices may be Euclidean, but the universe at large is not. To the question What happens to the mathematical *truth* of Euclidean geometry? Poincaré replies, 'It has no meaning ... One geometry cannot be more true than another; it can only be more convenient.'

Once learned, however, fundamental assumptions are difficult to transcend. Anyone trying to conceptualize the curvature of space unavoidably recognizes the strength of our Euclidean intuitions. This seems to be what don Juan refers to in Carlos Castaneda's *Tales of Power* when he says, 'The world doesn't yield to us directly, the description of the world stands in between.'(22)

The description of the world *does* stand in between. We create for ourselves a word-built world. We lock ourselves into this world to the extent that our thinking processes become dependent upon semantics. But we should not confuse our word-built reality with what actually is 'out there'.

This brings us to another important point revealed in the confluence of mysticism and the new physics. Not only do our fundamental assumptions inhibit us in our understanding of physics and metaphysics, but language itself becomes a hindrance. Both physics and metaphysics have reached a point where *language no longer imparts any information*. For instance, in quantum mechanics identical particles are said to be 'indistinguishable'. Two electrons that are indistinguishable can therefore be thought of as either the 'same' or 'different'. Our intuitions may tell us that they are different, but Heisenberg's uncertainty relations make this assumption meaningless. We therefore find that two words that are mutually exclusive become interchangeable. Neither imparts any information.

There is an alarming entropy to language which we have to watch out for. Metaphysics also presents many situations where language no longer imparts any information. In John Blofeld's translation of *The Zen Teaching of Hui Hai*, a disciple asks the Zen master: 'What is meant by perceiving the real Buddhakaya?'

Hui Hai replies, 'It means no longer perceiving anything as existing or not existing . . . Existence is a term used in contradistinction to nonexistence, while the latter is used in opposition to the former. Unless you begin by accepting the first concept as valid, the other cannot stand. Similarly, without the concept of nonexistence, how can that of existence have meaning? These two owe their being to mutual dependence and pertain to the realm of birth and death. It is just by avoiding such dual perception that we may come to behold the real Buddhakaya.' (14)

Once again, two words that are mutually exclusive become interchangeable. To the quantum physicist 'same' and 'different' no longer impart any information. To the Zen master 'existence' and 'nonexistence' no longer impart any information. This apparent dilemma should serve to point out the limitations of language. As the philosopher Wittgenstein observed, 'We are not analysing a phenomenon . . . but a concept . . . and therefore the use of a word.'

When the quantum physicist encounters indistinguishable electrons it is not the phenomenon that changes, but the fundamental assumption of how words function in our thinking processes. Whether it is the ultimate reality of the Buddhakaya or the subatomic realm, the mere fact that 'same' and 'different' no longer impart any information is a glittering insight into both the use of language and the phenomenon of electrons.

As will be shown in later chapters, the Everett-Wheeler interpretation of quantum theory states that the mathematical formalism of quantum physics yields its own interpretation. In simpler terms, all possible outcomes of an experiment exist, according to the interpretation, in an indefinite number of parallel realities. Aside from the dazzling implications of such a concept, the Everett-Wheeler metatheorem presents the same paradox encountered in the theologian's concept of an omnipotent God. If an experiment were ever performed to test the Everett-Wheeler metatheorem (the nature of such an experiment would be unfathomable) it would result in both proving and disproving it at one and the same time.

Similarly, in *God and Golem, Inc.*, the eminent cyberneticist Norbert Wiener states, 'I have already mentioned the intellectual difficulties arising out of notions of omnipotence, omniscience, and the like. These appear in their crudest form in the question often asked by the scoffer who turns up uninvited at religious meetings: "Can God make a stone so heavy that He cannot lift it?" If He cannot, there is a limit to His power, or at least there appears to be; and if He can, this seems to constitute a limitation to His power too.'(84)

Aside from being 'verbal quibbles' these two examples point out a difficulty centering upon the concept of infinity. In the nineteenth century, the mathematician Georg Cantor examined our intuitive notions of infinity and compared them with what was mathematically known about infinity. Cantor researched infinite sets and discovered that some infinities are 'larger' than other infinities. He created an entire system of mathematics around his astounding 'transfinite' numbers. Cantor demonstrated that there are as many even integers as there are integers. There are also as many integers as there are fractions. The number of these sets is the cardinal number *aleph null.*

Cantor also demonstrated that between any two points on a line there are 'more' than an infinite number of points; he referred to this greater than infinite number as aleph.

Between any two points within the line AB there are also aleph points, etc., ad transfinitum, which means that an aleph is equal to all of its parts. The number of points in a square is also aleph which is equal to the number of points in a cube which is equal to the number of points in an n-dimensional geometric solid!

Because aleph is equal to all of its parts the only way to obtain a number larger than aleph (a number that is greater than the number that is greater than infinity) is to raise aleph to the power of aleph. This number is aleph one and it has been demonstrated that aleph one is the number of all possible rational curves in space. Incredibly, Cantor revealed that it is possible to construct sets of higher and higher cardinality with no transfinite cardinal number as an upper bound. Indeed, the transfinite cardinals constitute a series whose end is unimaginable.(34)

Before we grow too upset at our incapacity to grasp transfinite numbers it should be noted that Cantor went mad playing with his alephs. The point, of course, is that the universe does not always represent itself to direct conceptualization. Both the Everett–Wheeler interpretation of quantum physics and the paradox of an omnipotent god demonstrate our incapability to comprehend infinity. The notion of infinity, infinity over zero, zero over zero, infinity raised to the infinite power, etc., are what Wiener calls 'indeterminate forms'. The difficulty they present lies fundamentally in the fact that infinity does not conform to the ordinary conditions of a quantity or number. Throughout this book we will find that both physicists and mystics are confronted with many indeterminate forms. Intuition betrays us, language fails us, and we will discover that our understanding of the universe depends upon modes of thought that Western civilization is only beginning to suspect. Our discovery of these indeterminate forms and our way of speaking and thinking about them is the exact point at which the confluence of mysticism and physics takes place.

We are led in classical physics to believe in a secure world of constructs and laws – a sense of space that tells us we can move in three dimensions and that the shortest distance between two points is a straight line; a sense of time that convinces us of the linearity of past, present, and future; a sense of causality that says every time I drop an object it will fall and every time I hit a billiard ball from the 'same' direction with the 'same' amount of force it will react in the 'same' way. The fact that we accept the cause and effect relationships between events as such an intimate part of our experience indicates once more the strength of our Euclidean intuitions. The implications of the new physics are just beginning to make themselves apparent.

As has been stated, the confluence of mysticism and the new physics calls such assumptions into serious question. In *The Interpretation of Nature and the Psyche*, Wolfgang Pauli and Carl Jung presented a *quaternio* which diagrams the basic concepts presently encountered in our understanding of the universe. As the authors stated, 'This schema satisfies on the one hand the postulates of modern physics, and on the other hand those of psychology.'(48) With the advent of the new physics, however, we will find that one by one, three of the concepts have fallen victim to a most peculiar maya. (Fig. 1)

In chapter one I examine the change of world view from observer to participant. It will be shown that the new physics has discovered

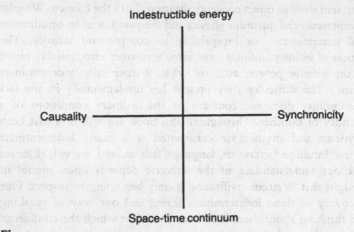

Fig. I.

that nothing even resembling causality exists on the atomic level. The implications of this are that the appearance of causality on the level of everyday life is simply a statistical quirk. The new physics suggests that the consciousness itself enters into the workings of the physical world and affects it. The implications of such a view are that there is no one reality. All possible realities coexist and some portion of the consciousness edits out all those realities which our intuitions cannot accept. The new physics suggests that the conscious-ness contains a 'reality-structurer', some neurophysiological mechan-ism which psychically affects reality itself.

In chapter two I examine what the new physics has to say about consciousness. Previously in scientific approaches the mere existence of consciousness as a phenomenon had been denied. In the new physics consciousness is found, perhaps, to be the only phenomenon that does indeed exist. Chapter two presents what the new physics has to say about the structure of consciousness and a mechanical explanation of the reality-structurer.

In chapter three the models of matter and space proposed by the new physics are examined. It will be shown that the new physics believes both matter and space owe their existence to the human consciousness. Reality, itself, is thus viewed as a 'super-hologram' which the consciousness creates for itself. The implications of this view are that the consciousness can enter in and alter the super-hologram to create changes in this reality.

In chapter four some of the views of the new physics concerning time are presented. It has been postulated in the new physics that there are regions which literally do not exist in either space or time. To conceptualize this the physicist has come up with an illustrative aid known as a light cone. Chapter four discusses the light cone and reveals some of the implications of a region which actually lies beyond space-time.

Chapter five is also on the nature of time in the new physics. The implications of the new physics are that the consciousness can actually work through the regions beyond space-time to effect phenomena which are normally thought of as impossible. In chapter five it is suggested that just as the reality-structuring portions of the consciousness can affect matter and space, they can also affect time. Again, it is suggested that the entire matter-space-time matrix of the physical universe owes its existence to the consciousness.

In the remaining chapters it will be shown that although these radical views are new in the realm of science, they are familiar to various branches of mysticism. In chapter six the remarkable similarities between an ancient branch of Hindu mysticism, or Tantra, and the new physics are clearly illustrated.

Thus, according to the new physics, there is no physical world 'out there'. Consciousness creates all. The purpose of chapter seven is to articulate the new world view proposed by the confluence of mysticism and the new physics. It is implied that there are no limits to the reality-structuring mechanisms of the human consciousness. Just as the mind can enter in and alter the super-hologram of reality, it can also create entirely new realities.

The purpose of chapter eight is to examine what the mystics have to say about the reality-structuring mechanism and how to achieve conscious awareness of it. It is implied that the reality-structuring mechanism is closely associated with the human nervous system and its control is effected by dealing with the human brain as if it were a computer or 'biocomputer'. Various yogas or methods of mind control are thus seen as little more than computer cards used to reach the reality-structuring portions of the human nervous system.

Finally, in chapter nine the new cosmology presented by the

confluence of mysticism and the new physics is clearly examined. If the implications of the confluence are that our assumptions about the physical universe need amending, then what assumptions are going to take their place?

It will be found that both mysticism and the new physics ultimately take the position articulated by Borges in *Other Inquisitions*. That is, we have dreamed the world. The full implications of this realization are just now beginning to make themselves apparent in the new physics. In the words of the astronomer Sir James Jeans, 'Today there is a wide measure of agreement, which on the physical side of science approaches almost to unanimity, that the stream of knowledge is heading towards a non-mechanical reality; the universe begins to look more like a great thought than like a great machine. Mind no longer appears as an accidental intruder into the realm of matter; we are beginning to suspect that we ought rather to hail it as the creator and governor of the realm of matter . . .'(44)

This is the melding of physics and mysticism. As we examine the universe closely we have to admit that the indeterminate forms we confront suggest an overwhelming maya about reality. It is the fabric of the dream that we arre touchinggWe have played with the notions of the idealists and now we must confront the observations of the physicists. How the omnijective nature of reality will change Western civilization remains to be seen. The only certainty is that the changes will be stupendous.

CONSCIOUSNESS AND REALITY

It must be admitted that the meaning of quantum physics, in spite of all its achievements, is not yet clarified as thoroughly as, for instance, the ideas underlying relativity theory. The relation of reality and observation is the central problem. We seem to need a deeper epistemological analysis of what constitutes an experiment, a measurement, and what sort of language is used to communicate its result. Is it that of classical physics, as Niels Bohr seems to think, or is it the 'natural language', in which everyone in the conduct of his daily life encounters the world, his fellow men, and himself? The analogy with Hilbert's mathematics, where the practical manipulation of concrete symbols rather than the data of some 'pure consciousness' serves as the essential extralogical basis, seems to suggest the latter. Does this mean that the development of modern mathematics and physics points in the same direction as the movement we observe in current philosophy, away from an idealistic toward an 'existential standpoint'?

HERMANN WEYL, *Philosophy, Mathematics and Natural Science*

CHAPTER ONE

Observer and Participant

Nothing is more important about the quantum principle than this, that it destroys the concept of the world as 'sitting out there', with the observer safely separated from it by a 20 centimeter slab of plate glass. Even to observe so minuscule an object as an electron, he must shatter the glass. He must reach in. He must install his chosen measuring equipment. It is up to him to decide whether he shall measure position or momentum. To install the equipment to measure the one prevents and excludes his installing the equipment to measure the other. Moreover, the measurement changes the state of the electron. The universe will never afterwards be the same. To describe what has happened, one has to cross out the old word 'observer' and put in its place the new word 'participator'. In some strange sense the universe is a participatory universe.

JOHN A. WHEELER, *The Physicist's Conception of Nature*

In 1927 Werner Heisenberg presented his famous Uncertainty Principle and sparked off a debate that has not yet been resolved. In simplified terms, Heisenberg stated that the observer alters the observed by the mere act of observation.(40) He was not implying that consciousness had any direct effect upon the outcome. He was referring instead to the problems encountered in trying to measure occurrences in atomic systems. Because of the incredible smallness of an atomic system, no observation can be made on a single system without seriously affecting it. This is roughly akin to saying that one cannot examine the machinery of a very small watch without disturbing its workings. The smallness of the system itself makes observation and measurement difficult.

The problem of observing atomic systems is increased by the fact that light can affect the system. For example, on the level of everyday life we take the act of observation for granted. Looking at a chair or a page of print seems to be a fairly aloof activity. The fact that the light reflecting off the chair and the page of print is actually altering them in minute ways is not immediately available to our perception. However, in very small systems such as the interior of an atom, a photon of light actually knocks the particles about. We can never be sure of the location of a particle because our only means of *seeing* the particle – bombarding it with a photon – will change its location (Fig. 2). We are like blind people trying to grasp at delicate spider webs.

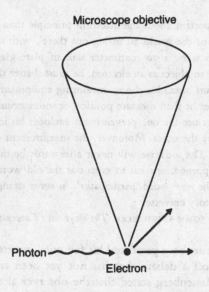

Microscope objective

Photon **Electron**

Fig. 2. *Gamma-ray microscope problem as discussed by Heisenberg in his uncertainty relations. The electron cannot be observed unless it interacts with a photon, in which case its recoil makes simultaneous momentum and position measurements impossible.*

It is up to the physicist to make atoms interact in systematic ways and, even after the observation has altered the system, try to deduce something about the atom's undisturbed properties. Heisenberg's Uncertainty Principle simply predicts the various possible effects observation will have upon the atom so that its undisturbed properties can be more closely estimated.

The original connection between the observer and the observed discovered by the quantum physicists therefore has to do with the technical difficulty of measuring atomic systems. From Heisenberg's findings there is no necessary indication that the consciousness of the observer affects the measurement, only the tools that the observer is forced to resort to, affect it. But later discoveries have caused some physicists to suggest that the human mind can affect matter.

One of the major revolutions in the realm of physics has been the increasing role of indeterminism – or the realization that it may be impossible to predict the outcome of an experiment no matter how much information we know about matter. Before the advent of quantum theory most physicists believed in a universe that was totally causal. In *Philosophical Essay on Probabilities* (1812–20), Laplace summed up such a position, stating, 'We ought then to regard the present state of the universe as the effect of its anterior state and as the cause of the one which is to follow. Given for one instant an intelligence which could comprehend all the forces of nature and the respective situation of the things that compose it ... for it nothing would be uncertain and the future, as the past, would be present to its eyes.'(27)

Classical physics seemed to show that causality existed on the level of everyday life. Everything from a weight oscillating on a spring to the motions of the planetary bodies represented systems that obeyed apparent laws of causality. Given the initial state of any system, all later states of the system could be predicted with great precision. The success of Newtonian physics is due to the fact that such laws seem to exist for virtually every system immediately perceivable to people, from billiard balls to computers, from electrical networks to eclipses. Where prediction was impractical, classical physicists still assumed that the system was causal. Thus, even if the physicist could not figure exactly where a bottle tossed into the Atlantic would end up, Laplace's hypothetical intelligence could.

On the level of quantum mechanical events, however, nothing even approaching causality has been found to exist. The well-known double-slit experiment provides an example that quantum theory contains concepts concerning matter which do not represent immediately observed quantities. Imagine a beam of particles all travelling with the same velocity. If the beam strikes a screen containing a narrow slit, the particles leaving the slit will no longer travel all in the same direction. They are *diffracted* so that their altered directions make various angles with respect to the initial beam. If the particles are then observed individually as they strike some detecting screen placed beyond the slit, they will not all strike within the area of the slit, but in a much wider area. Where an individual particle will be detected is neither predictable nor reproducible; only a pattern of distribution of hits can be formulated. For instance, a physicist might have one hundred particles waiting to be beamed through the slit. Known quantum mechanical laws enable the physicist to predict that 10 per cent of the particles will strike in one area and 90 per cent will strike in another. If the physicist allows one particle at a time to pass through the slit there is no way to predict which area the particle will choose. In fact, nothing can be found out to explain why a given 10 per cent of the particles strike one area and 90 per cent strike the other. The particles are identical. Strictly speaking, there is *no reason* why one particle should strike differently from another. In a universe that seems to be exceedingly causal such indeterminism creates a new world view whose repercussions have, perhaps, not even begun to jar the bedrock of classical physics.

The indeterminism of the quantum mechanical universe is surely one of the most startling changes in the way we view reality to occur in the realm of physics. It assaults our intuitions. It demolishes our misconceptions about the *connectedness* of events. There is little surprise in the fact that stanchions of the old school, including Einstein himself, opposed the denial of a causal universe. In *Einstein and Beckett*, Kuznetsov is quoted as observing, 'Einstein proceeds from the idea that a series of observable phenomena does not determine unequivocally the nature of the causal relationships between them. Hence the picture of causal relationships is to some degree deduced independently of direct observation. Einstein speaks of the free construction of concepts expressing causal relationships. Does this mean that such concepts are *a priori* or conceptual, or that causal concepts are arbitrary as a whole? The answer is No. The causal connection of processes may be expressed by means of different kinds of constructions, and in this sense their choice is arbitrary. But they must be in agreement with observation, and it is our duty to select the construction which agrees best.'(70)

It is important to understand that Einstein's specific objections to quantum theory were aimed at exposing its inability to give an adequate account of physical reality. To his death Einstein believed in a causal universe, one that simply doesn't exist on the level of atomic events.

In *The Human Use of Human Beings*, Norbert Wiener notes that it was neither Heisenberg nor Planck, but Willard Gibbs, who first proposed that the universe was contingent (predictable only within statistical limits) as opposed to deterministic. As early as the 1870s Gibbs was formulating his ideas on contingency. There is an overwhelming probability that every time you hit a billiard ball from the 'same' direction with the 'same' amount of force it will react in the 'same' way. But there are fringe occurrences – oddities on the frayed borders of our cause-effect reality – that indicate this contingency of the universe. In a contingent universe, although the billiard ball will react in the 'same' way most of the time, there is a chance that it will not react, or even do something totally unpredictable. According to Gibb's idea of contingency the physicist can no longer deal with what always happens, but only with what happens an overwhelming majority of the time.

Quantum theory also proposes that all systems can ultimately be described only statistically. The apparent causality of the universe is due to the fact that probabilities in systems larger than very small ones are very nearly equal to 1. The extremely high probability that eclipses will occur on certain determinable dates is the statistical result of an almost indefinite number of quantum mechanical events. Wiener states, '. . . in a probabilistic world we no longer deal with quantities and statements which concern a specific, real universe as a whole but ask instead questions which may find their answers in a large number of similar universes. Thus chance has been admitted, not merely as a mathematical tool for physics, but as part of its warp and weft.'(85)

It is this radical shift from a causal universe to a statistical one that created the most controversy. The implications of indeterminism are best revealed in problems first encountered by the Austrian physicist Erwin Schrödinger in the early part of the twentieth century. In quantum theory each variable governing a particle's behavior (energy, position, velocity, angular momentum, etc.) is taken into account. As physical experiments discover the properties and effect of these variables, the parallel task of the physicist is to formulate mathematical laws appropriately describing the physical properties and their relationships. Once a mathematical formalism has been created, the behavior of a particle can be predicted.

Take, for instance, the previous illustration of a beam of particles passing through a slit. Once the various properties of the appropriate operators are known, one can predict the pattern of distribution. Schrödinger developed the mathematical equation which describes this behavior. Because quantum particles exhibit complementarity – having the properties of both a particle and a wave – this equation is known as the wave function of the particle.

This is where the puzzle of indeterminism steps in. In certain circumstances Schrödinger's wave function predicts the behavior of a given particle up to a point and then describes two equally probable outcomes for the same particle. On paper as well as in observation *no reason* can be found for the particle's varying behavior. The question therefore seems to have entered a kind of schizophrenic state in which it cannot decide which outcome to choose. Quantum theory does not deal with individual events. Given an individual particle, Schrödinger's wave function cannot determine where it will strike the screen, but can only predict where an ensemble or group of particles will strike.

The implications of this indeterminism are even more dramatically illustrated in an interesting thought problem popularly known as Schrödinger's cat. The set-up is as follows: A cat is trapped in a room containing a Geiger counter and enough radioactive material so that in one hour there is a 50 per cent chance one of the nuclei will decay. Upon discharge of the counter a specially attached hammer will break a flask of poisonous gas. According to Schrödinger's wave function at the end of the hour the system will have a form in which the living cat and the dead cat are 'mixed' in equal portions (Fig. 3). Naturally, if this is performed in experiment, only one observable outcome will occur, leading Schrödinger to feel that the mathematics create a paradoxical and unacceptable description of reality.(30)

In 'Time and Quantum Theory' (1966), J. Zimmerman states, 'These questions of individual events are, in the language of conservative quantum theory, meaningless; and therefore, the interpretation goes, they are in fact meaningless. Only questions and statements about ensembles are meaningful.'(90) As might be expected, such an indeterministic view of the universe was opposed by the classical physicists. John von Neumann first suggested that Schrödinger's equation might be in error, in 1955. He introduced a second equation to ascertain the error but that, too, became schizophrenic. So did a third equation, and a fourth, etc., creating a chain known as 'von Neumann's catastrophe of infinite regression.'(30)

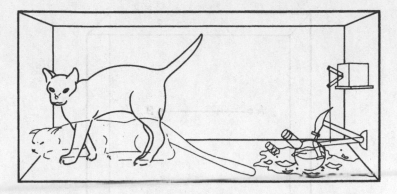

Fig. 3. *Schrödinger's cat. At the end of the experiment the equation predicts that the cat will be both* living *and* dead *in mixed proportions*
(from B. S. DeWitt, 'Quantum Mechanics and Reality', Physics Today, 23, no. 9 (1970): (30).

The indeterministic nature of atomic systems was not easily accepted by most physicists. It is roughly similar to a table of billiard balls where some balls react when they are struck, some remain motionless for several minutes before they move, and some move without being struck. To more easily conceptualize the paradoxical description of reality represented in Schrödinger's equation, imagine the wave function as an abstract function in configuration space, an imaginary three-dimensional space used for conceptualizing problems. The behavior of the particle can be represented as a line in the configuration space (Fig. 4).

At the point where Schrödinger's equation predicts two equally probable outcomes the line branches (Fig. 5a). According to the wave function, the single particle performs two different behaviors at one and the same time. Under certain conditions the wave function will predict an infinite number of schizophrenias, in which case its path (or vector) in configuration space branches into four possible outcomes, eight possible outcomes (Fig. 5b), sixteen possible outcomes, *ad infinitum*.

Since a cat that is both living and dead in mixed proportions, as well as a particle that performs two or even an infinite number of possible behaviors at one and the same time, are totally contrary to our experience, Schrödinger's equation doesn't seem to be describing individual

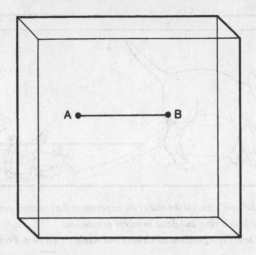

Fig. 4. A *is the particle's initial state;* B *is its state as it progresses through time (described by Schrödinger's equation).*

atomic events. Various interpretations of quantum theory have been offered to account for the schizophrenia of Schrödinger's equation.

The conventional interpretation chosen by most quantum physicists is known as the 'Copenhagen collapse'. According to this view, when the equation divides into two, one of the vectors in configuration space simply collapses. Instead of consisting of a multitude of outcomes the equation reduces to a single result. Proponents of the Copenhagen collapse argue that quantum theory is strictly indeterministic. The equation does not represent reality, but is only an algorithm (a mathematical method) for making statistical predictions. For instance, if Schrödinger's cat experiment is performed one obviously does not end up with a cat that is both living and dead in mixed proportions. Only an ensemble of living and dead cats would present an accurate description of reality.

Michael Audi states, '. . . if indeterminism is genuinely accepted, all the philosophical problems of interpreting quantum theory become tractable'.(4) Opponents argue that the assignment of statistical weights and the arbitrary collapse of the vector do not follow from Schrödinger's equation. Einstein and de Broglie contended that a rigid deterministic world was more acceptable than a

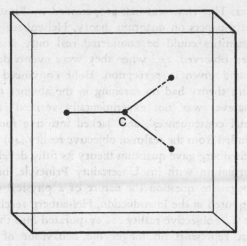

Fig. 5a. *At C Schrödinger's wave function contracts schizophrenia and its vector in configuration space divides as the particle is predicted to perform two mutually exclusive outcomes.*

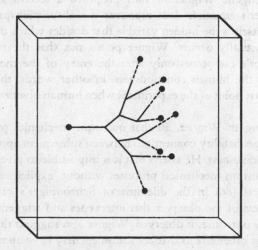

Fig. 5b.

contingent one and suggested another reason for the schizophrenia of the equation. Perhaps all information concerning quantum mechanical events was not known and an undiscovered or 'hidden variable' might be responsible for the varying behavior of the two particles. This view was first propounded by Einstein in 1935.

In his early papers on quantum theory, Heisenberg insisted that physical quantities could be considered real only after they had actually been observed, i.e., when they were events describable in space-time and given in perception. Bohr convinced Heisenberg that quantum theory had no meaning in the absence of a classical realm. Whatever was 'not experimentally verified', revealed no 'observational consequences', and 'lacked intuitive foundation' he simply excluded from the realm of objective reality.(40)

After Heisenberg gave quantum theory its fully developed mathematical formalism with his Uncertainty Principle, he and other physicists began to question the nature of a physical or objective reality. As quoted in the Introduction, Heisenberg reached the conclusion that '. . . objective reality . . . evaporated into the . . . mathematics that represents no longer the behaviour of elementary particles but rather our knowledge of this behaviour'.(41)

In an attempt to maintain the existence of an objective reality and still resolve the puzzle of the wave function, Nobel Prize winning physicist Eugene Wigner in 1961 proposed a second solution. If Schrödinger's equation does represent a reality, perhaps the consciousness itself is the hidden variable that decides which outcome of an event actually occurs. Wigner points out that the paradox of Schrödinger's cat occurs only after the entry of the measurement signal into the human consciousness. In other words, the paradox occurs at that point of the experiment when human observation intervenes.

According to Wigner, all that quantum mechanics purports to provide is probability connections between subsequent apperceptions of the consciousness. He asserts that it is impossible to give a description of quantum mechanical processes without 'explicit reference to consciousness'.(86) In the dilemma of Schrödinger's cat it is the consciousness of the observer that intervenes and triggers which of the possible outcomes is observed. Wigner also suggests that a search be made for other effects that consciousness may have upon matter.

In *Symmetries and Reflections* Wigner sketches a possible mathematical description of what he believes must occur when consciousness affects the observation. He states, 'The preceding argument for the difference in the roles of inanimate observation tools and observers with a consciousness – hence for a violation of physical laws where consciousness plays a role – is entirely cogent so long as one accepts the tenets of orthodox quantum mechanics in all their consequences. Its weakness for providing a specific effect of the consciousness on matter lies in its total reliance on these tenets – a reliance which would be, on the basis of our experience with the ephemeral nature of physical theories, difficult to justify fully.'(86)

The idea that consciousness affects matter is an unusual statement for a physicist. In its mechanistic and empirical approach science has always striven to exorcise the ghost of consciousness from any formulation of the laws of physics. Wigner's suggestion that the relationship between consciousness and objective reality, like the nature of causality, needs to be reexamined is a radical departure from classical physics. Even though Wigner proposes a new relationship between the observer and the observed, he maintains that the line between consciousness and reality 'cannot be eliminated'.(86) There are still two kinds of reality – subjective and objective. The classical realm of objective reality simply becomes relative.

Princeton physicist John A. Wheeler believes that the term 'observer' should be replaced by the term 'participator'. This replacement, he feels, would explicitly point out the radical new role of consciousness in physics. Instead of denying the existence of objective reality he further asserts that subjective and objective reality sort of *create each other*. They are 'self-excited systems' and are brought into being by 'self-reference'. As he puts it, 'May the universe in some strange sense be "brought into being" by the participation of those who participate? . . . the vital act is the participation. "Participator" is the incontrovertible new concept given by quantum mechanics. It strikes down the term "observer" of classical theory, the man who stands safely behind the thick glass wall and watches what goes on without taking part.' As Wheeler concludes, 'It can't be done, quantum mechanics says.'(80)

Wheeler's suggestion of the term 'participator' demonstrates the mystical nature of the new physics. We may recall Sir James Jeans' assertion that the mind may be the creator and governor of the realm of matter. Similarly, in an article entitled 'Implications of Meta-Physics for Psychoenergetic Systems', physicist Jack Sarfatti also asserts his belief that the *structure of matter may not be independent of consciousness!*(67)

Sarfatti further proposes that we have to incorporate a logical calculus of two-valued 'yes-no' propositions to fully understand quantum theory. This yes-*and*-no logic leads us to the third interpretation of quantum mechanics, the Everett-Wheeler or 'many worlds' interpretation, which expresses a view of the universe common to science fiction, but contrary to the intuitions of most physicists. The interpretation proposes that the universe is continually splitting into a stupendous number of parallel realities. In s ch a universe we not only exist in an indefinite number of worlds, but all possible outcomes of any event also exist.

In his short story 'The Garden of the Forking Paths', Jorge Luis Borges tells of a mythical Chinese nobleman named Ts'ui Pên, who, during his life, vows to do two things: write a book and construct a maze. It is only after he dies that his descendants realize the two projects are one and the same. The book, *The Garden of the Forking Paths*, is cryptic and apparently irrational. In the first chapter the main character is killed. In the second he is alive again. Every time one of the characters is faced with several alternatives, he or she chooses all of them simultaneously.

Borges describes a protagonist who finally realizes the vision concealed in *The Garden of the Forking Paths*. It is a theoretical work on the nature of time: '. . . a picture, incomplete yet not false, of the universe as Ts'ui Pên conceived it to be. Differing from Newton and Schopenhauer . . . [he] did not think of time as absolute and uniform. He believed in an infinite series of times, in a dizzily growing, ever spreading network of diverging, converging and parallel times. This web of time – the strands of which approach one another, bifurcate, intersect or ignore each other through the centuries – embraces every possibility.' As the protagonist explains, 'We do not exist in most of them. In some you exist and not I, while in others I do, and you do not, and in yet others both of us exist. In this one, in which chance has favored me, you have come to my gate. In another, you, crossing the garden have found me dead. In yet another, I say these words, but am an error, a phantom.'(17)

Although Borges' work is fiction, Ts'ui Pên's conception of time parallels the Everett-Wheeler interpretation of quantum mechanics. Basically three problems haunt the various interpretations of quantum mechanics. First, von Neumann's attempt to check Schrödinger's equation for errors assumes that the mathematics is incorrect. The wave function describes a reality contrary to intuition, but, as the catastrophe of infinite regression reveals, the mathematics have never been proven in error. Second, the collapse of the equation employed by the Copenhagen school avoids any explanation for such a statistical phenomenon. Last, proposals such as Wigner's assume the existence of a physical reality even though Heisenberg's findings make the definition of such a physical reality impossible.

In 1957 Hugh Everett, along with John A. Wheeler, examined the issues. They subsequently created the Everett-Wheeler interpretation of quantum mechanics, which requires no changes in the basic mathematics of the Schrödinger equation. In its basic premises it:

1. Accepts the mathematics of Schrödinger's equation.
2. Accepts that none of the branches of Schrödinger's equation collapses.
3. Denies the existence of a physical reality.

The Everett-Wheeler hypothesis accepts the conventional probability interpretation of quantum theory by making a important distinction. Probability as it relates to quantum theory is *different* in concept and should not be confused with probability as it is understood in statistical mechanisms. Quantum theory mathematically describes a universe in which chance is not a measure of our ignorance about a system, but is absolute. It is inevitable that states such as the schizophrenia of the wave function should occur. The branches of the wave function separate and divide according to the various possibilities of a given measurement. The behavior is part of the mathematics of the Schrödinger equation. Because chance is not a measure of our ignorance of the system, the new information should not make us deny or alter the equation.

The problem, of course, is the same one confronted by Ts'ui Pên's descendants. The Everett-Wheeler interpretation accepts the three problems encountered by other interpretations, but challenges our intuitive conceptions of time. As Bryce DeWitt points out, Everett and Wheeler propose a universe that '. . . is constantly splitting into a stupendous number of branches, all resulting from the measurement like interactions between its myriads of components. Moreover, every quantum transition taking place on every star, in every galaxy, in every remote corner of the universe is splitting our local world on earth into myriads of copies of itself.'(30)

The possibility of 10^{100+} universes, all imperfect copies of each other, and all totally unaware of one another's presence, has awesome implications. In Schrödinger's experiment, for every cat that survives in our universe, in another universe one dies. The wave function causes the universe to split in two and the paradox is resolved. As von Neumann's catastrophe of infinite regression implies, every quantum mechanical event in our universe causes an indefinite number of divisions in which probability dictates that all possible realities 'exist'. In such a garden of the forking paths the solution to the dilemma of indeterminism may be a universe in which all possible outcomes of an experiment actually occur (Fig. 6).

This interpretation, like all those before it, has its problems. A mathematical formalism that yields its own interpretation can never receive operational support in the laboratory. Even if such a test experiment could be performed, like the von Neumann test apparatus, it would become schizophrenic and again yield all possible outcomes. The Everett-Wheeler interpretation, like a cat that is both living and dead in mixed proportions, would prove itself and disprove itself at one and the same time.

However, just as we are forced to accept the yes-*and*-no logic in the fact that electrons appear to be waves and particles at one and the same time, perhaps we should heed Wheeler's (and Sarfatti's) suggestion of a yes-*and*-no calculus. The Everett-Wheeler interpretation (whether ontologically correct or not) is perhaps the only suitable answer.

This brings us once again to the relationship between consciousness and reality. Wheeler's 'participator' is implicitly suggested even in the many worlds interpretation. If both outcomes of Schrödinger's cat experiment *do* occur, some triggering device in the human consciousness must *decide* which outcome to experience. Jack Sarfatti believes it is the unconscious minds of the collective human race that determines whether a particle decays or not. However, because we are unaware that we are participators in the quantum universe, our collective will is unfocused and incoherent and this is what makes quantum events seem so random and probabilistic.(67)

Sarfatti further postulates that the concept of the participator can be employed to explain other phenomena as well. For instance, in Brownian movement, or the constant zigzag movement of particles

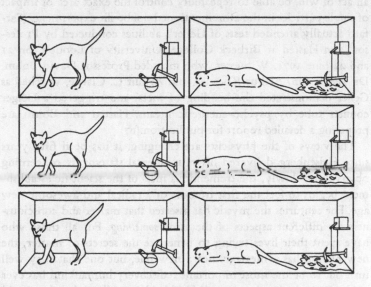

Fig. 6. *All possible outcomes of an experiment occur in an indefinite number of universes*
(from B. S. DeWitt, 'Quantum Mechanics and Reality', Physics Today, 23, no. 9 (1970): (30).

in a liquid or gas, he proposes that the mind of the participator determines the movement of the particles. The random character of Brownian movement is due to the assumption that the collective will of the participators is generally unfocused. Thus, Sarfatti suggests, the particle is buffeted about in a random Brownian motion created by the subconscious mental functioning of all participators. He states, 'The participator in a particular quantum experiment in a physics laboratory can be the experimenter himself, though on the deeper level of quantum interconnectedness it must also include the general range of all living systems. All conscious systems, independently of their *spatio-temporal* locations relative to the experimental apparatus, make incoherent contributions to the total nonlocal quantum potential felt by the individual photons or electrons.'(67)

Sarfatti feels that the participator principle is responsible for the apparent miraculous talents demonstrated by psychics such as Uri Geller. If such talents are valid, individuals such as Geller should, by an act of will, be able to repeatedly control the exact area of impact of single particles in the aforementioned doubt-slit experiment. (Sarfatti actually attended tests of Geller's abilities conducted by Professor John Hasted at Birbeck College, University of London, on 21 and 22 June 1974. Witnesses, who included Professor David Bohm, Dr Ted Bastin, Arthur Koestler, and Arthur C. Clarke, watched as Geller demonstrated the bending of metal and triggered a Geiger counter tube by psycho-energetic means. Hasted and Bohm are preparing a detailed report for publication.(67)

The views of the physicists are changing. It has been fifty years since Heisenberg delivered his monumental statements concerning observation; slowly, the tremendous mass of the scientific establishment begins to feel the first tremors of a radical and awesome new age. For centuries the mystic has asserted that matter and consciousness are different aspects of the same *something*. For all those who have spent their lives trying to penetrate the secrets of matter, the new physics has a message, not a new one, but one that may well turn out to be the most important rediscovery humankind has ever made. Perhaps the change will be felt like a roll of thunder as old constructions fall and new ones take their place. Perhaps the change will be so subtle and gradual that we will have no more sense of it than the anti-Copernicans during the time of Galileo, who did not feel the earth move. Whatever the case, the message of the new physics is that we are *participators* in a universe of ever-increasing wonder. We have penetrated matter and found a bit of maya – and a glimpse of ourselves.

CHAPTER TWO

A Holographic Model of Consciousness

It now seems highly plausible that the 'seat of consciousness' will never be found by a neurosurgeon because it appears to involve not so much an organ, or organs, but the interaction of energy fields within the brain. These patterns of energy would be disrupted by surgical intervention, and have long since disappeared in cadavers. Neurophysiologists will not likely find what they are looking for outside their own consciousness, for that which they are looking for is that which is looking.

KEITH FLOYD, '*Of Time and the Mind*'

Holograms are three-dimensional images created with the aid of a laser. For instance, when you shine a laser beam through a piece of photographic film containing the encoded image of an apple, a three-dimensional image of the apple will appear on the other side of the film. The most intriguing thing about a hologram is that if you cut the film in half and shine a laser through each piece, two complete three-dimensional images of the apple will appear. Cut the film into four pieces and four apples appear. Cut it into eight and eight images appear, and so on. This is because each piece of a holographic transparency contains the entire image.*

The property of being 'holographic' or having every part contained in the whole is remarkable because it indicates that the organization of the information contained in a hologram is much different from the organization of information in normal pictures. In an impressionist painting, for example, each bit of information or daub of paint can be isolated and viewed separately from the painting as a whole. However, a hologram cannot be divided up into fragments. Because each apparent bit of a holographic image can only be understood as it relates to the collective bits of the

---

*This effect does not work with holograms in which the image is visible to the naked eye so do not cut one of these in half or you will ruin your hologram.

entire picture, we may speak of it as possessing certain 'field' properties.

The holographic model of organization is pertinent to our discussion because the new physics has found that the fundamental units of matter (subatomic particles) also cannot be isolated as individual bits or building blocks. Their behavior also possesses field properties in that it is determined by the collective of particles. This is intriguing because the same holographic/field relationship also appears to govern the structure of life and, indeed, the structure of our thinking processes as well.

As the quote by Floyd (above) suggests, we may expect the riddle of consciousness to be a unique problem for several reasons. First, consciousness itself is the only tool we have to examine consciousness. As such, we are locked into an endless dilemma – a mirror infinitely reflecting a mirror situation, from which there is no escape. Such a situation is analogous to the problems of making measurements considered by Heisenberg. The observer alters the observed. The thinker alters the thought.

Of course the point of view that a large portion of the scientific community has held for most of the twentieth century is that consciousness does not exist. Wittgenstein's argument that the existence of the subjective 'I' should be removed from our language because no physical measurements can be made upon consciousness has become an almost passionate conviction of many scientists.

In *Design for a Brain* (1952) pioneer cyberneticist W. Ross Ashby has written an entire book on the mathematics and organization of a thinking machine without a single reference to consciousness. Ashby does not specifically deny the existence of consciousness, but feels simply that learning as a process has no 'necessary dependence' on consciousness. The behaviorists 'and their allies', as Arthur Koestler puts it, are leading contemporary proponents that consciousness, as such, does not merit study by the scientific community.(49)

But their passionate conviction is waning. With the advent of the new physics more and more scientists are reexamining the puzzle of consciousness. In a paper titled 'The Confluence of Psychiatry and Mysticism', psychiatrist Stanley R. Dean lists a compendium of mystical beliefs that are gaining wider scientific acceptance. Among them he articulates the hypothesis that thought has universal 'field' properties which, like gravitational and magnetic fields, are 'amenable to scientific research'.(29)

In our approach to a holographic/field model the first question is: What processes in the brain are associated with the consciousness? It has been firmly established that an electric polarization exists across the membrane which separates the interior nerves of the brain's neural web from the surrounding medium. The membrane has the ability to decrease polarization by altering its permeability to certain ions. Thus, when a nervous impulse is initiated, the depolarization travels along the membrane of a nerve cell until it reaches the point of junction between adjacent neurons, or a synapse. At this point a potential difference exists across the synaptic cleft. In certain circumstances the transmission across the synaptic cleft cancels out another simultaneously arriving impulse; in others, the transmission creates another impulse to fire the neuron.

It is unknown at present what process is involved in the transfer. The problem of consciousness therefore becomes: *What process is involved in the interconnection of all portions of the brain?* There is clear evidence that the process is not chemical or electrochemical. Evan Harris Walker of the NASA Electronics Research Center, Cambridge, Massachusetts, postulates that 'consciousness is a nonphysical but real quantity'. He argues that the process involved is not necessarily chemical and suggests that the transfer may be due to a 'quantum mechanical tunneling' process. He offers convincing evidence that some sort of quantum mechanical phenomenon takes place at the synapse, but admits that such a process does not completely explain consciousness: 'We need to find some process that allows transitions over long distances (several centimeters) and yet does not conflict with what is known about synapses and preferably does not alter the process which we have already described as taking place at the synapse.'(77)

If the riddle of consciousness is the interconnection between all portions of the brain, there appear to be two possible ways of explaining such an interconnection. The first is that taken by Walker, i.e., the connection is effected by means of some particle-reaction, either chemical, electrochemical, or quantum physical. The second approach is that the interconnection is effected by means of a force field extending over the appropriate region of space.

One possibility is an electromagnetic field. However, because the currents in the brain do not propagate along the length of the dendrites and axons, the various regions of the brain do not appear to be electrically interconnected. The electrical fluctuations of the brain are, in fact, quite local and involve only the immediate vicinity of a small portion of any given nerve cell. Unless further experimentation proves otherwise, we can dismiss electromagnetic fields as the process. But what, then, does create the interconnection?

Our answer, perhaps, rests in a similar problem encountered by the quantum physicists. Some physicists believe that an interconnection or 'quantum potential' exists between atomic particles, but, as in the neurophysiology of the brain, no interconnecting field or process has been found. For instance, one such example of interconnectedness occurs in the previously mentioned double-slit experiment. Imagine that we start with 100 subatomic particles. Say that Schrödinger's equation has enabled us to predict that 10 per cent of these particles will strike in area A and the remaining 90 per cent will strike in area B. As has already been stated, the behavior of an individual particle cannot be predicted. Only the pattern of distribution of the entire group of particles follows predictable statistical laws. If we let the particles pass through the slit one by one we will notice that after 10 per cent of the particles have struck area A, further particles passing through the slit seem to *know* that the probability has been fulfilled and shun the area.

D. Bohm and B. Hiley state, 'It would follow that somehow the measurement of the momentum of the first particle actually "put" this particle into a definite state of momentum p_1 while it "put" the second particle into a correspondingly definite correlated state of momentum $p - p_1$. The paradoxical feature of this experiment is that particle 2 somehow seems to "know" into which state it should go, *without any interaction that could transmit information.*'(16)

Bohm and Hiley have postulated that the particles that strike area A must somehow be interconnected with the remaining particles in order to 'know' that one of the probabilistic niches has been filled. They propose that this interconnectedness is due to the existence of a subtle and as yet undetected field they call the 'quantum potential'. However, they do not believe that the particles should be thought of as communicating with one another, at least not in the conventional sense of sending signals back and forth to one another. As they state, 'The mere fact of interaction does not necessarily give rise to the possibility of carrying a signal. Indeed, a signal has, in general, to be a complex structure, consisting of many events that are ordered in definite ways.'(16) Rather Bohm and Hiley believe that the interconnections between the particles are acausal, and that the quantum potential field allows them to behave as if they were actually one entity despite their apparent separateness in space and time.

There are many striking similarities between the quantum potential and the interconnection of the human brain. Both deal with levels of organization in which the behavior of discrete entities, synapses or subatomic particles, seems to be governed by the collective. Both appear at first glance to involve signaling processes that facilitate a transfer of information, and yet all known means for sending signals fall short of explaining all the interconnections that seem to exist in each system. Is there, then, some chance that the two respective processes are related? Some researchers think so.

The major obstacle in creating a model of consciousness involves a misconception basic to both neurophysiology and quantum physics. It concerns the shift in the scientific world view from 'causality' to a more holographic or 'holistic' approach. In classical physics the holistic aspects of a system were virtually ignored. In 1968 the father of General Systems Theory Ludwig von Bertalanffy stated, 'The only goal of science appeared to be analytical, i.e., the splitting up of reality into ever smaller units and the isolation of individual causal trains. Thus, physical reality was split up into mass points or atoms, the living organism into cells, behavior into reflexes, perceptions into punctual sensations, etc. Correspondingly, causality was essentially one-way; one sun attracts one planet in Newtonian mechanics, one gene in the fertilized ovum produces such and such inherited characteristic, one sort of bacterium produces this or that disease, mental elements are lined up, like the heads in a string of pearls, by the law of associations.'(9)

Understandably, a holistic view of nature is very slow in coming. Only since the advent of quantum theory has the scheme of isolable units acting in one-way causality proved unsatisfactory. Von Bertalanffy points out that notions of teleology and directiveness are being considered by physicists hesitantly because they previously appeared to be outside the scope of science. He further observes that such notions were commonly held 'to be the playground of mysterious, supernatural or anthropomorphic agencies . . .'(9) Once again science and mysticism meet at the crossroads.

In the field of biology the transition is slowly occurring. Classical scientists have always maintained that the DNA code contains all the information needed to shape and organize biological systems. In *The Fields of Life*, Dr Harold Saxton Burr of Yale University School of Medicine suggests a more holistic approach and presents an excellent study of the electrodynamic fields which organize the structure of living entities or L-fields.

Another analogy can be drawn between the quantum potential and biological organisms. In a developing salamander embryo modern biology is currently at a loss to explain how the longitudinal axis of the tail is established. At early stages of development the cells which will ultimately become part of the tail show no apparent specialization. If one experimentally rearranges the cells, even though their ultimate fate is altered, the longitudinal axis does not change. Cephalic cells become caudal cells; right cells become left cells – with little interference in the process of growth. In some way the constituent cells of the growing system have their state determined and their behavior and orientation controlled. Although no interaction occurs that could transmit information, and the cells are originally indistinguishable (as indistinguishable, one might say, as the particles in the double-slit experiment), they pattern themselves in an apparent collective behavior.

The German biologist Hans Driesch long ago pointed out that the fate of any group of cells in an embryo is not only genetically conditioned, but is also the result of the position of that group of cells in the biological whole. Driesch called the mechanism by which position could determine cellular potencies 'entelechy' and described an extra-biological guiding principle much like the quantum potential.

Burr's discovery of L-fields both fulfills Driesch's description of entelechy and provides a working holistic mechanism to explain the collective behavior of cells. As Burr states, 'The following theory may . . . be formulated. The pattern or organization of any biological system is established by a complex electrodynamic field which is in part determined by its atomic physio-chemical components and which in part determines the behavior and orientation of those components. This field is electrical in the physical sense and by its properties relates the entities of the biological system in a characteristic pattern and is itself, in part, a result of the existence of those entities. It determines and is determined by the components.'(20)

As Burr points out, the electrodynamic field has holographic properties in that any portion of the L-field contains the entire design of the organism. In the early stages of its development if the cells of an embryo are divided in half, each half will develop into a completely formed organism. You will not end up with the front half and the back half of the same organism. Just as the division of a hologram results in two complete images, the division of the developing embryo results in two identical twins. Like the hologram it appears that every minute part of the L-field contains the blueprint for the whole.

Keith Floyd proposes that a 'holographic model of consciousness' makes such brain processes as memory, perception, and imaging clearly explainable. In consciousness, one frame is every frame. Every memory and bit of information stored in our minds is infinitely cross-referenced with every other bit of information in a 'creative pattern of pure and perfect ambiguity'. Floyd suggests that the 'screen' of awareness may thus be viewed as an organic form of a holographic plate which processes three-dimensional perceptions and reconstructed images with equal facility.(36)

This is, perhaps, the most incredible thing about consciousness. If the organic holograph does not have three-dimensional perceptions to process, it creates its own reality to perceive/conceive. Individuals placed in sensory deprivation chambers begin to hallucinate and synthesize entire *inner* realities. When the human mind is cut off from the so-called physical world it has the remarkable ability to create its own world – trees, people, sounds, colors, and smells – in what John C. Lilly refers to as 'cognitional multi-dimensional projection spaces'.(54) The entire physical universe itself is nothing more than patterns of neuronal energy firing off inside our heads!

In terms of what is going on in the neurophysiology of the mind there is no difference between cognitional multi-dimensional projection spaces and what we perceive as being external reality. In fact, since there are only 100 million sensory receptors, and about 10 trillion synapses in the nervous system, the consciousness is about 100,000 times as receptive to changes in its internal environment as to its external environment. Not only is the universe inside our heads equal to the physical universe in terms of the neurophysiological processes involved – but the internal environment may be more real. This may well be the ultimate demonstration that all worlds are in the mind.

In *The Human Biocomputer*, John C. Lilly describes various experiences and observations he gained while in total sensory deprivation. As he explains, when the consciousness is cut off from all external stimuli the human mind begins to create its own environment. *'One is aware of "the silence" in the hearing sphere; this too gives way to the new space which is developing, as fear or other need builds up. As with the "darkness and the silence" so with the presence or absence of the body image.'*(52) Slowly, Lilly's inner self began to feel more at ease with the vast spaces within. As he explains, his mind began to explore the projection spaces. First it created simple environments, a single sound, a solitary image. But as his consciousness grew more accustomed to the new territory the projections became more and more complex. Ultimately, Lilly was able to experience entire 'other universes' in his cognitional multi-dimensional projection spaces.

Implicit in the mind's ability to totally reconstruct its environment is the fact that the brain has an incredible ability to store information. As Pieter Van Heerden points out, if the brain only stored one bit (unit) of information per second for a lifetime it would require an incomprehensible 3×10^{10} elementary binary nerve impulse operations to accomplish this.(76) Amazingly, the brain's capabilities can store many more bits than one per second; again, only a holographic model of consciousness appears to explain such a talent. Photographic holograms possess a fantastic capacity to (retrievably) store information. Image after image can be superimposed upon a holographic plate simply by varying the wave length of the light. Each image retains its identity and can be recovered without affecting the other images. Indeed, some 10 billion bits of information have been successfully stored holographically in a cubic centimeter! Van Heerden suggests that a holographic process might explain the brain's similar capability.

Similarly, neurophysiologist Karl H. Pribram hypothesizes a holographic model of consciousness. Holographic representations are incredible associative mechanisms. They have the ability, Pribram points out, to instantaneously perform cross-correlations – the very properties attributed to thought in the problem solving process. As Pribram suggests, 'Holograms are the "catalysts of thought". Though they remain unchanged, they enter into and facilitate the thought process.'(65)

There are universes inside our heads – universes superimposed upon universes. A passage from the Upanishads illustrates the point beautifully. 'When one goes to sleep, he takes along the material of this all-containing world, himself tears it apart, himself builds it up, and dreams by his own brightness, by his own light. Then this person becomes self-illuminated. There are no chariots there, no spans, no roads. But he projects from himself chariots, spans, roads. There are no blisses there, no pleasures, no delights. But he projects from himself blisses, pleasures, delights. There are no tanks there, no lotus-pools, no streams. But he projects from himself tanks, lotus-pools, streams. For he is a creator.'

Interestingly, Keith Floyd points out that some neurophysiologists are coming to regard higher brain functions in terms of an optical system processing a form of bioluminescence. This light inside the skull may be the very self-illumination that the Upanishads refer to. Floyd further proposes that the area of the midbrain immediately posterior to the optic chiasma is the locus of the neural holographic plate. The pituitary gland, thalamus, hypothalamus, and pineal body in particular appear to be associated in the theater of conscious awareness. The pineal body is thought by many to be a vestigial sensory organ and is partly composed of light-sensitive tissue similar to that found in the retina of the eye. This, Floyd asserts, seems 'to lend support to the speculation that it might serve as the "grid" of patterned ambiguity on which perceptions are constructed and memories are reconstructed'.(36) How appropriate, considering that the pea-sized organ has long been regarded in the East as the 'third eye' or mystical doorway to spiritual awareness.

In attempting to determine the relationship between the various organs of the brain, neurophysiologists run into an interesting problem. If the pineal body does play a primary role in memory and perception, its removal should produce profound and even total disruption of these functions. However, this is not the case. The excision of the pineal body in rats disrupts the biological clock of the organism, but apparently very little else. As Floyd concludes, 'Further reflections on the progress suggested that the "screen", the holographic plate which I had so long been attempting to identify with an organ, may actually be a function of an area instead of an organ. It began to appear that the pineal body occupies the midpoint at the center of a neural energy field, at which point occurs the burst of light that is experienced as the screen of consciousness on which shifting figure-ground relationships represent external reality.' Floyd thus concludes that consciousness does not involve so much an organ, or group of organs, but the interaction of energy fields within the brain.(36)

The interconnection of all portions of the brain in a screen of conscious awareness reveals striking field properties similar to both the relationship between cephalic cells and caudal cells in a developing salamander embryo and the relationship between two indistinguishable particles in the double-slit experiment. We might suspect that here is the secret to the relationship between mind and matter. As Dr Burr puts it, 'In the last analysis, the Universe is a unit, all of its parts are related to the wholeness of the Universe, and there is necessarily some interrelationship between the wholeness of the Universe and the activities of its individual components. From the unified theory of Einstein – even though it lacked final, complete validation with respect to the law of gravity – it is clear that one of the characteristics of the Universe is fields which can be measured by instruments. It does not make any difference whether you call it an electro-static field, an electro-magnetic field, or an electro-dynamic field. The name is always a consequence of the methods which were applied to its study. In other words, there is one unifying characteristic of the Universe which we have ignored, and that is its field properties.'(20)

Physicists Bohm and Hiley point out the same holistic aspects governing the organization of matter that Burr postulates in his L-fields. They state, 'Any attempt to assert the independent existence of a part would deny this unbroken wholeness . . . This does *not* necessarily mean that the subsystems are always spatially smaller (localized) than the system as a whole. Rather, what characterizes a subsystem is only its relative stability and the possibility of its independence of behavior in the limited context under discussion.'(16)

The switch from a causal universe to a holistic one, the asserted interdependence of all parts comprising the whole, Burr's speculation that gravity may be a master field which governs the organization of all phenomenon – all are views that are totally contrary to our notions of classical physics. In 1968 philosopher Geoffrey Chew formulated such a radical departure in world view into what he calls a 'bootstrap' philosophy. The universe seems to be lifting itself up by the bootstraps, so to speak. The bootstrap philosophy constitutes the final rejection of a mechanistic world view such as that proposed by Newton. No longer can we look at the world as built of fundamental entities with fundamental properties. The universe cannot be understood as an assemblage of independent parts like the daubs of paint in an impressionist painting. It is a hologram, a dynamic web of interrelated events in which each part of the web determines the structure of the whole.(21)

A holographic view of consciousness (and, indeed, a holographic view of the entire universe) is, perhaps, the closest physics can come to mysticism without the two losing their identities. One is reminded of the metaphor of Indra's net mentioned in the *Avatamsaka Sutra* in which the universe is viewed as a sort of cosmic network of interpenetrating things and events. In the words of Sir Charles Eliot, 'In the heaven of Indra, there is said to be a network of pearls, so arranged that if you look at one you see all the others reflected in it. In the same way each object in the world is not merely itself but involves every other object and in fact *is* everything else. "In every particle of dust, there are present Buddhas without number." '(32)

The value of a holographic view of consciousness can be summarized as follows: First, a holographic view of consciousness displaces the view of the behaviorists that all our mental behavior can be interpreted in terms of stimulus and response. Our thought processes are holographic in that all thoughts are infinitely cross-referenced with all other thoughts. The storage of information in the brain is an incredibly complex process. We cannot view it as an alphabetical file; otherwise every time someone mentions the word 'ocean', for instance, we would have to ponderously backtrack through all the associations the word 'ocean' ever contained for us. But we find that we do not have to look through some immense file in a time consuming search. Somehow, the word 'ocean' instantly brushes against all of our thoughts and memories simultaneously to bring out the associations we search for. This is the key to creativity, that every thought is contained in every other thought like the pearls in Indra's net, or like Buddhas contained within Buddhas contained within Buddhas. Our thoughts are like Chinese dolls; each thought contains every other thought, and only a holographic view of consciousness can provide an adequate metaphor for such an unfathomable process.

The second aspect of a holographic view of consciousness is the implications it has for consciousness as a field. If consciousness is a field and only one vibration on the continuum of fields that organize matter, we have an explanation for the interaction between mind and matter. Physicist Jack Sarfatti articulates this view and, like Burr, hypothesizes that gravitation is the unified master field of the universe, and is responsible for the interaction between consciousness and matter. Physicists currently believe that gravity is mediated by a particle called a 'graviton' and it is actually the graviton that is responsible for the large-scale structure of galaxies and the universe at large. Sarfatti embraces this belief and proposes that living systems may similarly be organized by 'biogravitons' (of which there may be several subvarieties). Consciousness can influence and control these biogravitons which in turn interact with the fields that govern the structure of matter.(68)

The possibility of psychokinesis immediately follows from Sarfatti's postulate that consciousness controls the biogravitational field. This field, the biogravitational field of consciousness, may interact with all the other resonances (levels of organization) such as the gravitational field of Einstein, and even the atomic and nuclear fields which govern the structure of matter. Such a cosmic bootstrap picture of mind and reality provides a physical explanation for what we might call the reality-structurer – that portion of the consciousness which enables certain individuals to overcome the alleged laws of physics.

The field of consciousness may be on the same continuum as the field which gives the illusion of the quantum potential. The interaction of both fields would most assuredly explain how the mind of the participator can affect where a particle in the double-slit experiment will strike. The hologram of consciousness is a biogravitational field and the hologram of matter is a gravitational field (more will be said on the holograph of matter in the next chapter). Matter and consciousness are a continuum. Burr's statement that the L-field determines and is determined by its components and Wheeler's proposition that the universe is created by the participation of those who participate are observations on the holistic aspects of reality itself. In this light the mind and the universe become one immense cognitional multi-dimensional projection space – or simply fields, within fields, within fields.

PART TWO

THE STRUCTURE OF SPACE-TIME

The lack of definiteness which, from the point of view of empirical importance adheres to the notion of time in classical mechanics, was veiled by the axiomatic representations of space and time as things given independent of the senses. Such use of notions — independent of the empirical basis to which they owe their existence — does not necessarily damage science. One may, however, easily be led into the error of believing that these notions, whose origin is forgotten, are necessary and unalterable accompaniments of our thinking, and this error may constitute a serious danger to the progress of science.

ALBERT EINSTEIN, *Out of My Later Years*

CHAPTER THREE

Superspace

A tree, a table, a cloud, a stone – all are resolved by twentieth-century science into one similarly constituted thing: a congeries of whirling particle-waves obeying the laws of quantum physics. That is, all the objects we can observe are three-dimensional images formed of standing and moving waves by electromagnetic and nuclear process. All the objects of our world are 3D images formed thus electromagnetically – super-hologram images if you will.

CHARLES MUSES, *Consciousness and Reality*

On 19 July 1967, former director of the Foreign Economic Administration and assistant to the Secretary of Commerce Arthur Paul visited a small fishing village in Tamil, seventy miles north of Colombo, Ceylon (now Sri Lanka). He was invited by the local member of Parliament to witness the traditional yearly fire-walking ceremony.

The spectacle took place in front of an old Hindu temple. Before the temple a pit fifteen feet long and five feet wide had been prepared and filled with burning coals. The people in charge of building the pit continuously doused themselves with water to keep cool enough to work close to the fire. As Paul reports in his journal, he sat twenty or thirty feet from the pit and could still feel its intense heat. Paul continued to explain that the fire walkers prepared themselves with a ceremonial bath in the sea about a mile from the temple. They waited until a crowd of three hundred people gathered. As the excitement of the crowd heightened, the high priest, in striking headdress, stepped into the pit.

Slowly and sedately he walked across the live coals, revealing not the slightest sign of pain. Behind him followed fifteen or twenty of the more faithful natives of Tamil. Some carried babies in their arms. Some carried children. Two young boys jumped out of the red-hot coals, unable to make the passage, but two little girls, not more than eleven or twelve years old, walked the entire distance unharmed. A number of the walkers returned for a second and even a third crossing. In the final part of the annual ritual the high priest performed his customary 'faint' and was carried to the temple. The ceremony was over.

In his journal, Paul stated, 'The path leading to the pit had become wet from the dousing of the workers who prepared the burning coals. Therefore the feet of the workers may have picked up some damp mud before they crossed the pit and there may have been some sand on their feet from the sea. But the coals were so intensely hot that this does not seem to me to be an explanation for the ability of these people to perform this feat; in fact, I could see no ordinary rational explanation for what I had watched.'(57)

The phenomenon of fire walking remains one of the most documented and enigmatic examples of how *consciousness affects reality*. All attempts to explain it in the framework of classical physics have so far proved unsuccessful, and only underline the phenomenon as an incongruity, a glistening unreality in the maya of our classical conception of space-time and causality.

At Surrey, England, in 1935, the English Society for Psychical Research performed a series of tests on two Indian fakirs. Both physicians and psychologists from Oxford attended to witness their fire-walking abilities. The fakirs demonstrated that they were able to withstand red-hot coals whose surface temperature was 450°–500° Celsius (842°–932° Fahrenheit) and whose interior temperatures reached 1400° Celsius (2552° Fahrenheit). They repeated their fire walk with no chemicals, no preparations, and no harm whatsoever to their flesh.

At the close of their amazing performance one of the fakirs informed one of the incredulous psychologists that he too could fire-walk unharmed if he would simply hold the fakir's hand as he did so. The psychologist bravely shed his shoes and hand in hand the two walked the fire unharmed.(38)

Within the classical framework phenomena such as fire walking remain unexplained. Even accepting that all systems are contingent as opposed to causal, a peculiar state of affairs remains. If the laws of physics are statistical quirks, one can *uneasily* accept that fire walking is one of those one-in-a-million occurrences. Under a strict view of cause and effect, fire burns. Under a contingent view of cause and effect, there is an extremely high probability that fire burns, and fire walking is a freak roll of the dice. But the fact that the freak roll of the dice occurs annually in Tamil is a direct indication that some human agency is involved in the statistics.

As any fifth grader knows, fire or heat is a form of energy whose effect is produced by the accelerated vibration of molecules. If the human consciousness participates along with matter it is not premature to hypothesize that the consciousness of the high priest of Tamil somehow intervenes in the accelerated vibration of the molecules and arrests the normal fire-burns process. This reality-structuring ability is very similar to Sarfatti's suggestion that the random behavior of particles in Brownian motion may be linked with the volitional activity of the experimenter. As was previously stated, a possible explanation for the reality-structurer is that the consciousness can generate a biogravitational field which can interact and alter the gravitational field governing matter. The purpose of this chapter is to examine the structure of matter proposed by the new physics in an attempt to understand the extent to which consciousness might be able to affect the physical universe.

The mechanistic view of the universe proposed by Newtonian physics was based on the notion that reality is composed of basically two things: solid objects and empty space. In the realm of everyday life this notion is still valid. The concepts of empty space and solid material bodies are a basic part of our way of thinking about and dealing with the physical world. The realm of everyday life may thus be viewed as a 'zone of middle dimensions', or the realm of our daily experience in which classical physics continues to be a useful theory. Consequently it is very difficult for us to imagine a picture of reality where the concepts of solid objects and empty space lose their meaning. However, empty space has lost its meaning in light of the findings of Einstein and the concept of solid objects has been virtually destroyed by the investigations of quantum theory.

Since as early as the time of the Greek atomists, as demonstrated in the writings of Democritus and Leucippus, the classical notion of matter has been that it is composed of basic building blocks or atoms. Just what these atoms are became the holy grail of the physicist. If one divides a mountain one will find that it is composed of many rocks. If one divides a rock one will find that it is composed of many grains of sand. Logically, if one keeps on dividing the grains of sand one should ultimately obtain one of these basic building blocks. This was the reasoning of the physicists.

It was not until the turn of this century, however, that science got its first peek into the structure of the atom. With the discovery of X-rays a virtual window was created into the world of the very small. Soon other radiations were discovered, such as the radiations emitted by radioactive substances. The phenomenon of radioactivity provided positive proof that atoms are not the basic building blocks of matter, but composites, built of entities even smaller. It was up to the scientist to figure out ways to utilize the newly discovered radiations as further tools to pry open the secrets of the atom. Physicists such as Max von Laue used X-rays to explore the arrangements of atoms in crystals; Ernest Rutherford found that tiny particles emitted by radioactive substances, or alpha particles, were as helpful as the doctor's scalpel in dissecting the building blocks of matter.

When Rutherford attacked the atom with high-speed alpha particles he found something totally unexpected. Far from being the solid and physical particles they were believed to be since the fifth century B.C., atoms turned out to consist of vast empty regions of space in which incredibly tiny particles — electrons — orbit around a nucleus. Indeed, atoms are so incredibly small that it is difficult for us to grasp their structure without resorting to mental illustrations. For instance, if a baseball were expanded to the size of the earth the atoms that comprise it would be seen as being roughly the size of cherries. It would look as if trillions upon trillions of cherries had been packed into a mammoth sphere. Remarkably, if we had an atom the size of a cherry we still would not be able to see its nucleus with the naked eye. If we blew the atom up to the size of a basketball or even large enough to fill a room, the nucleus would still remain too small to be seen. If we continued to increase the size of the atom until it was the size of the dome of St Peter's Cathedral in Rome (one of the largest domes in the world), the nucleus would only have the size of a grain of salt.

The discovery that matter is mainly composed of empty space is only the first of many discoveries to destroy the physicist's notions of solid objects on the atomic level. With the advent of Heisenberg's Uncertainty Principle and quantum theory the solid world of matter could never look the same. Rutherford's experiments revealed that

matter consists mainly of vast empty regions of space. The findings
of Heisenberg and the quantum physicists were that the building
blocks of the atoms themselves – electrons, protons, neutrons, and a
host of other subatomic particles – did not even display the properties of other physical objects. Subatomic units of matter simply do
not behave like solid particles. Instead, they appear to be abstract
entities.

Depending upon how we look at it, a subatomic entity displays
the properties of both a particle and a wave (Fig. 7). When these

Wave   Particle

Fig. 7.

entities behave like particles they act as if they are packed into a
very small volume of space, somewhat like buckshot. When they
behave like waves they appear to be spread out over large regions
of space. What is the physicist to do? Can the fundamental units of
the atom be *something* that neither our language nor our conceptions
can deal with? Heisenberg vividly portrayed the shock of the physicist in the following recollection: 'I remember discussions with
Bohr which went through many hours till very late at night and
ended almost in despair; and when at the end of the discussion I
went alone for a walk in the neighbouring park I repeated to myself
again and again the question: Can nature possibly be so absurd as it
seemed to us in these atomic experiments?'(41)

In an attempt to understand a reality which does not easily
accommodate words, Heisenberg proposed that the physicist should
simply accept the complementarity or paradoxical aspect of subatomic entities, and view them as wave/particles – entities which can
only be explained by concepts which are interrelated and cannot be
defined simultaneously in a precise way. By doing so Heisenberg
was making a statement that belonged as much to mysticism as it
did to the new physics. That is, the ultimate nature of reality is
beyond verbal description. The greatest commonality in both mysticism and the new physics is that both point to the inadequacy of
language.

Time and again the limits of our language define the limits of our understanding of the universe. The indistinguishability of subatomic particles points out still another weakness of our linguistic approach to understanding reality. For instance, all electrons are exactly alike. It is as if they are all mirror reflections of one another – there is no way to distinguish between two electrons. To the physicists it therefore becomes as meaningful to say that two electrons are the 'same' as that two electrons are 'different'.

In his *Lettres à Maurice Solvine* (Paris, 1956), Einstein clearly points out the problem: 'Concepts can never be regarded as logical derivatives of sense impression. But didactic and heuristic objectives make such a notion inevitable. Moral: it is impossible to get anywhere without sinning against reason: in other words, one cannot build a house or a bridge without the use of a scaffolding which, of course, is not part of the structure.'(68) The Zen Buddhists say that a finger is needed to point at the moon, but we should not trouble ourselves with the finger once the moon is recognized. The finger is a scaffolding for understanding the notion of the moon. The words 'particle' and 'wave', and 'same' and 'different', are also scaffoldings for understanding the nature of matter. We may suspect that all words are scaffoldings and in this sense try to transcend their limits. Such are necessary sins against reason.

It has grown increasingly difficult for both the physicist and the mystic to be aware of the limits of conceptual knowledge. In fact, this seems to largely be the purpose of such crystalline branches of mysticism as Zen and Taoism – not so much to impart information as to reveal the limits of our symbolic methods of dealing with information. In other words, we should always recognize that our representation of reality may be easier to grasp than reality itself, but we should not confuse the two. This seems to be what the Taoist sage Chuang Tzu meant: 'Fishing baskets are employed to catch fish; but when the fish are got, the men forget the baskets; snares are employed to catch hares; but when the hares are got, men forget the snares. Words are employed to convey ideas; but when the ideas are grasped, men forget the words.'(75) And Heisenberg: 'Every word or concept, clear as it may seem to be, has only a limited range of applicability.'(41)

Towards the end of the nineteenth century Max Planck discovered a phenomenon similar to the complementarity of subatomic particles. He found that the energy of heat radiation was not emitted continuously, but appeared in the form of discrete units or energy packets. Einstein called these units of radiation 'quanta' and thus gave quantum theory its name. Einstein further suggested that all forms of radiation, including light, can be propagated in the form of either waves or quanta. Indeed, it was soon discovered that light behaved much like particles or 'photons' and was propagated in discontinuous quanta. Unlike electrons, however, photons are massless and always travel at the speed of light.

The similar dual nature of particle and wave in the electron leads to curious problems when we try to examine the elusive little creatures. When we say we know where an electron is in an atom, this isn't the same as knowing where a planet is around the sun. In order to follow a classical orbit, the electron would, at any given instance, have to have a definite value for its exact position and its exact velocity at the same time. However, as Heisenberg's uncertainty relation points out, any measurement taken in order to follow the orbit of the electron would disturb it so much that we would not be able to determine what orbit it was following. There is no way of telling whether electrons follow orbits in the atom, and no reason to believe they do. Once again, this is a difficulty in measurement. However, even when we think we've pinpointed an electron's position exactly, it might fool us and be someplace else entirely. This is because electrons don't behave solely as particles. Sometimes they seem to be localized in a small region of space and sometimes they are spread over a much wider region. (Fig. 8) Electrons don't exist as objects exist. They only reveal 'tendencies to exist' and even when we've taken a very accurate measurement of the position of an electron, it only means there is a high probability of finding the electron there. This is not a difficulty in measurement, but inherent in the nature of the electron. Because electrons possess both the properties of a particle and a wave packet, they cannot be said to have distinct geographic locations. An electron cannot be held like a leaf or a seashell. No physicist will ever 'see' an electron or touch it, for the electron is a phenomenon which our concepts and our language cannot pin down. Not only is the universe queerer than we think, but it is queerer than we *can* think.

Is there any hope then that we can figure out what quanta are?

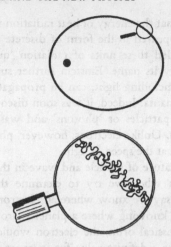

Fig. 8. *A close-up of an electron as it might appear 'spread out' over space (from John A. Wheeler, 'Superspace and the Nature of Quantum Geometrodynamics' in C. De Witt and J. A. Wheeler,* Battelle Rencontres, 1967 Lectures in Mathematics and Physics, *W. A. Benjamin: New York, 1968).*

If they are the fundamental building blocks of matter, surely something can be determined about their nature. To this question the new physics replies yes – we can accept subatomic entities as being abstract entities, and we can say something about their nature. To do so, however, we must bring in perhaps the most difficult concept of all, the concept of curvature of space.

In 1916 Einstein presented his general theory of relativity. In it he asserted that space is not three-dimensional and time is not a separate entity. Space and time are different aspects of the same *something*, according to Einstein. They comprise a four-dimensional continuum in which there is no universal flow of time as in the Newtonian view of the universe. Different observers will order events differently in time. How they order their view of time is relative to their positions and velocities in relation to the observed events. Einstein's greatest contribution to our view of the universe was his assertion that all measurements involving space and time are not absolute. In the general theory of relativity the classical notion of space as the stage of physical events is abandoned. Both space and time become elements of a language that any given observer employs in a description of the universe.

Even more startling was Einstein's hypothesis that space is not Euclidean, but curved. According to Einstein the very fabric of space-time possesses a geometric property or curvature which reveals itself in such phenomena as gravity. We may conceptualize gravity as curvature by imagining that masses such as the earth are resting on vast sheets of rubber. Such a mass would sink into the rubber much as the earth actually 'sinks' into the fabric of space-time (Fig. 9). Event entities without mass are affected by the curvature; thus

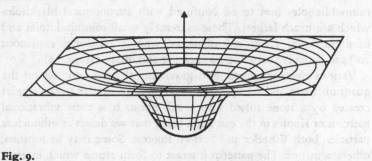

Fig. 9.

light rays are seen to 'dip' closer to large planetary bodies as they pass by them. The curvature of space-time is difficult to conceptualize without thinking of space-time as having some substance. But it is the nothingness itself that is curved. As Sir Edmund Whittaker observed, 'In Einstein's conception space is no longer the stage on which the drama of physics is performed: it is itself one of the performers; for gravitation, which is a physical property, is entirely controlled by curvature, which is a geometrical property of space.'(83)

It is the substance of this nothingness that some physicists believe is the true basic building block of matter. As early as 1876 W. K. Clifford was theorizing that matter is nothing more than empty, curved space.(38) John A. Wheeler epitomizes Clifford's views when he states, 'there is nothing in the world except empty curved space. Matter, charge, electromagnetism, and other fields are only manifestations of the bending of space. *Physics is geometry*.'(79)

To Wheeler the nothingness of space can be seen as composed of fundamental building blocks. If we could examine it microscopically we would find that the fabric of space-time or 'superspace' is composed of a turbulent sea of bubbles. These bubbles are the warp and weft of empty space and comprise what Wheeler calls the 'quantum foam'. He states, 'The space of quantum geometrodynamics can be compared to a carpet of foam spread over a slowly undulating landscape . . . The continual microscopic changes in the carpet of foam as new bubbles appear and old ones disappear symbolize the quantum fluctuations in the geometry.'(79)

Jack Sarfatti has elaborated upon Wheeler's imagery. He imagines the quantum foam as a turbulent sea of rotating miniblackholes and miniwhiteholes (not to be confused with astronomical blackholes which are much larger). These extremely small miniblackholes and miniwhiteholes (10^{-33} cm diameter with a relatively enormous 10^{-5} gm mass) are constantly appearing and disappearing.(68)

Various electromagnetic and gravitational forces can act on the quantum foam and set up vibrational patterns similar to the ripples created by a stone tossed into a still pond. It is these vibrational patterns or ripples in the quantum foam that we detect as subnuclear particles, both Wheeler and Sarfatti suggest. Some may be protons, others neutrons. The patterns interact to form atoms which interact to form molecules which interact to form the substance of the physical world. Thus, in some odd way, the stones and the stars are merely undulations in the nothingness!

This, of course, is only a crude conceptualization of curvature or what curved space-time may be like. It is difficult for us to conceive of curved space. Our thought processes are locked into our concept of our reality; it is a near impossibility to imagine the curvature of space-time, or twists and bends in the material of reality itself. We would have to look at the fabric of this reality from outside of reality, and this we cannot do at present. We are caught in the funhouse mirror and the light that brings us images of the physical universe follows the curves of the mirror's surface. The distortions are invisible to us.

The concept of matter as undulations in the quantum foam may shed light on the paradox of complementarity. As has been stated, one of the basic puzzles facing the physicist is that subnuclear entities such as electrons and protons exhibit the properties of both waves

and particles. If Wheeler's suggestions are correct, attempts to measure an electron are roughly akin to trying to measure the ripple on
a streamer of silk flowing in the wind.

All along, the physicist has searched for the ultimate substance of
matter and viewed empty space as simply the stage of the material
world. Atomic particles were conceived as being something quite
different from this empty stage, but as Whittaker suggests, with the
advent of the new physics the stage itself becomes one of the actors.
The wave/particle duality of matter is our first hint that matter and
so-called empty space are much more intimately related. As Wheeler
stated, 'Is space-time only an arena within which fields and particles
move about as "physical" and "foreign" entities? Or is the four-
dimensional continuum all there is? Is curved empty geometry a
kind of magic building material out of which everything in the
physical world is made: (i) slow curvature in one region of space
describes a gravitational field; (ii) a rippled geometry with a different
type of curvature somewhere else describes an electromagnetic field;
(iii) a knotted-up region of high curvature describes a concentration
of charge and mass-energy that moves like a particle? Are fields and
particles foreign entities immersed *in* geometry, or are they nothing
but geometry?' (79)

In the new physics, matter and empty space-time thus become
one and the same. The basic building blocks are not objects in the
sense that we know them, but may, as Sarfatti suggests, be viewed
as tiny miniblackholes and miniwhiteholes. Wheeler proposes that
these bubbles in the quantum foam resemble 'teacup handles' or
'wormholes' through the fabric of space-time. These wormholes
can connect two different regions in space much as a hollow teacup
handle connects two areas inside a cup (Fig. 10).

The distance between the mouths of two wormholes as we might
perceive them in three-dimensional space and via the route through
the hollow teacup handle can be very different, Wheeler points out.
For instance, a wormhole in San Francisco and a wormhole in New
York are thousands of miles apart in three-dimensional space, but
the distance through the teacup handle that connects them might be
only a few inches. Thus the wormholes are actually holes in space.
Because the bubbles in the quantum foam are constantly being
created and destroyed, space 'resonates' between one foamlike
structure and another. The apparent three dimensions of space are

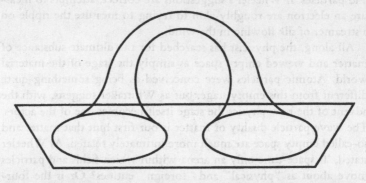

Fig. 10. *The 'wormhole' connects two regions in Euclidean space. The distance through the two mouths of the wormhole and the distance through three-dimensional space may be of entirely different magnitudes (as illustrated to a very limited degree).*

virtually nonexistent on the level of the quantum foam. The worm-holes of Wheeler's superspace create a 'quantum interconnectedness' in which every point in space is connected to every other point in space.

Many writers have pointed out that the existence of such worm-holes in space-time forces us to view every geography and history in the universe as a 'nonlocality', or directly adjacent to every other geography and history. D. Bohm and B. Hiley state, 'It is generally acknowledged that the quantum theory has many strikingly novel features . . . However, there has been too little emphasis on what is, in our view, the most fundamentally different new feature of all, i.e., the intimate interconnection of different systems that are not in spatial contact.'(16)

All things are interconnected. Assertions by George Berkeley and Alfred North Whitehead that consciousness and the physical world are connected gain new significance in the light of Wheeler's proposi-tion. Not only are time travel and faster than light travel a real possibility, but we now must suspect that every point in the human brain is connected, via the quantum foam, to every other point in the universe. Often this omnijective connection between the mind and the universe is likened to a dream reality. As has been stated, in a dream the division between the consciousness and reality are arbitrary. I can dream that I and several friends are sitting in chairs and talking. But the division between myself, the chairs, and my dream image friends is only an illusion. All artifacts and entities are subordinate to the consciousness of the dreamer. The dream reality is ultimately omnijective.

In a universe whose warp and weft is the quantum foam, the fabric of physical reality becomes indistinguishable from the fabric of a dream. Wheeler's superspace casts doubt on the strict three-dimensionality of things. Because the wormholes connect every point in space with every other point the universe collapses into a peculiar one-dimensionality. In effect, from a perspective beyond the quantum foam, or literally beyond space-time, the universe would seem to have no dimensionality at all. A similar situation is encountered in dream reality. We may dream of vast spaces, of three-dimensional rooms with chairs, tables, and people. But the dimensionality of the dream has no presence outside the dreamer.

Sir James Jeans' notion of the universe as a giant thought as opposed to a giant machine is further echoed by the quantum physicist. Jack Sarfatti states, 'Signals move through the constantly appearing and disappearing (virtual) wormhole connections, providing instant communication between all parts of space. These signals can be likened to pulses of nerve cells of a great cosmic brain that permeates all parts of space. This is a point of view motivated by Einstein's general relativity in the form of geometrodynamics. A parallel point of view is given in the quantum theory as interpreted by Bohm. In my opinion, this is no accident because I suspect that general relativity and quantum theory are simply two complementary aspects of a deeper theory that will involve a kind of cosmic consciousness as the key concept.'(68)

The omnijective continuum of a dream provides an adequate explanation for a causal phenomenon. For instance, I can dream that I gaze at a flower and cause it to bloom. In a dream this is perfectly normal. I do not wonder why the flower bloomed in accordance with my whim. No psychokinetic energies or interaction need pass between the flower and myself to explain this cause and effect. The consciousness of the dreamer generates the space-time of the dream.

In a slightly more bizarre vein I might dream that I am an electron in the double-slit experiment. I am one of the 10 per cent to pass through the slit and strike the screen in that area. Because a statistical law of distribution that is also my whim has been fulfilled,

I *will* that all further electrons passing through the slit strike other areas of the screen. Again, no electromagnetic energies or interaction need pass between the electron (myself) and the other electrons in order to explain the quantum potential that exists between us.

Bohm and Hiley's work suggests that we cannot analyze the universe by dividing it up into parts. As they state, 'Our work brings out in an intuitive way just how and why a quantum many-body system cannot properly be analyzed into independently existent parts with fixed and determinate dynamical relationships between each of the parts. Rather, the "parts" are seen to be in immediate connection, in which their dynamical relationships depend, in an irreducible way, on the state of the whole system (and indeed on that of broader systems in which they are contained, extending ultimately and in principle to the entire universe). Thus, one is led to a new notion of *unbroken wholeness* which denies the classical idea of analyzability of the world into separately and independently existent parts . . .'(16)

We might suspect that a dreamlike nature would once again entail such an unbroken wholeness. That the relationship between any two particles depends on something going beyond what can be described in terms of the particles alone, indicates that the universe may be holographic. Much like the holographic model of consciousness presented in the last chapter, Sarfatti's cosmic consciousness, Sir James Jeans' giant thought, Bohm and Hiley's unbroken wholeness provide us with a picture of the universe that the mystics have been pointing out for centuries.

We are told, then, that if the reality we perceive as the physical universe is examined microscopically, we will find it to be a super-hologram. As Charles Muses puts it, 'We live in a projection world of solid, neuro-"wired" holograms – a world of simulacra.'(57) In truth, the leaf and the mountain are only configurations of microscopic, turbulent wave/particles.

It is easy for us to accept that the image on the television set is really a contrivance: a mirage shaped by various electromagnetic energies. It is more difficult for us to imagine reality as a similar mirage: gravitationally trapped light locked into super-holograms, and that is all. As Muses observes, '. . . here we demur, saying, but that is not "all" at all; for that scene existed or exists somewhere and the projection apparatus is, if anything, the more illusory of the two – certainly not the ultimate reality of the scene it serves to depict. Indeed, the entire science of the projection is irrelevant to the scene's reality.'

The world seems oppressively solid. In trying to figure out the *substance* of reality we reach the impasse and ironic security that being locked in the funhouse mirror affords. With a particularly Pythagorean approach, Wheeler finds that the fabric of space-time, the substance of reality itself, is nothing but geometry. In 'The Space-Time Code', David Finkelstein asserts that space-time is a statistical construct from a deeper 'pregeometric' quantum structure in which *process* is fundamental: 'According to relativity, the world is a collection of *processes* (events) with an unexpectedly unified causal or chronological structure. Then an object is secondary; is a long causal sequence of processes, a world line. According to quantum mechanics the world is a collection of *objects* (particles) with an unexpectedly unified logical or class-theoretical structure. Then a process is secondary; is a mapping of the objects or of their initial to their final conditions.'(35)

Finkelstein asks, 'Which are we to build out of our quanta – beings or becomings, essences or existences?' Finkelstein follows what he calls an existential physics and favors the process model. Reality is process. The physical world is process. Any attempt to penetrate the ultimate meaning of substance of this process brings us again up against a most peculiar maya – pure geometry, undulations in the nothingness.

We may suspect, like Muses, that a more important issue concerns the projection apparatus of the super-hologram. Finkelstein proposes that primitive systems, say, an electron, are elementary processes and not objects at all. Such primitive processes are assembled into 'chromosome-like code sequences' to build simple objects, which are braided and cross-linked to make more complex objects and their interactions. Should we ask the question: Are the chromosome-like code sequences part of the projection apparatus?

The performance of the high priest of Tamil is evidence that our minds may generate some force or field which affects this space-time code. Just as an electromagnetic radiation can distort the picture on the television set, the consciousness can distort the chromosome-like code sequences we know as fire and affect the super-hologram of reality.

The image flickers.

The super-hologram is altered ever so slightly and a process that we have come to expect as absolute, that fire burns, becomes a process of a much more startling nature. The priest walks the coals unharmed.

CHAPTER FOUR

Beyond the Light Cone

Space and time are not concepts which can be meaningfully applied to single microscopic systems. Such systems are to be described by abstract concepts (charge, spin, mass, strangeness, quantum numbers) which make no reference to space and time. These microscopic systems interact in ways that must also be described abstractly, that is without reference to space and time. When a vast number of such microscopic systems so interact, the simplest and most fundamental result is the creation of a space-time framework which gives validity to the classical notions of space and time, but on the macroscopic level only.

J. ZIMMERMAN *'The Macroscopic Nature of Space-Time'*

According to the Tantric mystics of Tibet, our perceptions of a universe existing in time are incorrect. Above and beyond this illusory reality is the void – a region where the concept of time itself ceases to have any meaning. The Buddhists also recognize a world which exists beyond time. As eminent Zen scholar D. T. Suzuki states, 'In this spiritual world there are no time divisions such as the past, present and future; for they have contracted themselves into a single moment of the present where life quivers in its true sense . . .'(25)

Because of our apparently linear and sequential experiencing of past, present, and future, it is hardly surprising that we interpret time as an absolute as opposed to a construct. But the physicists are slowly destroying this last myth and are developing an approach to time which more closely resembles the view long held by the mystics. At the moment we are caught between the future and the past in the immeasurable interim of the present. Nothing ever happens in the past (or the future). Everything occurs in the present. These are things we assume without question. So when the physicist Richard Feynman suggests that a positron moving forward in time is actually an electron moving backwards through time, we must pause. Our thinking cannot readily accommodate the possibility that part of our universe (and even part of our consciousness) might exist beyond the prison of time.

The purpose of this chapter is to compare our classical notions of time to the views of the quantum physicists and the mystics. It is suggested that the nature of time is quite different from what we have been led to expect. If the views of the quantum physicist should prove to be true, what implications does this have for the realm of everyday life? What becomes of the shape of time?

In Newtonian physics, space and time were viewed as independent entities. Time was thought of as being absolute for all parts of the universe. According to this way of thinking a clock in New York (assuming that it is an ideal clock and never loses or gains time) will tick away the same number of seconds as an ideal clock in Moscow, or an ideal clock in the Andromeda galaxy.

According to Einstein's special theory of relativity, time is relative to its frame of reference. To specify a time measure, according to Einstein, we must specify the motion of the observer relative to the proper frame of reference. For instance, an observer holding a clock can count his or her heartbeats and directly calculate a pulse rate (heartbeats per minute). However, the same observer when in motion relative to the clock cannot measure the time interval properly. He may derive a different pulse rate because the clock appears to run at a different rate from when he was at rest. Both frames of reference, the observer and the moving clock, have nonabsolute time systems relative to themselves.

Einstein proposed, as did Minkowski before him, that space and time are not separate entities. They are a continuum, or different aspects of the same fundamental 'something'. Their ultimate interchangeability is like that of matter and energy. According to the theory of relativity they are merely the elements of a language that expresses the laws of nature in one frame of reference or another. In configuration space, space-time can be thought of as a four-dimensional coordinate system with the three spatial dimensions as the first three axes and time as the fourth axis. For any event E we can draw a diagram (Fig. 11), where the x-axis represents a direction in space and the ct-axis represents time measured in units compatible with those of x. The axes represent the space-time coordinates for the given event E. For another observer moving relative to the first, the coordinates

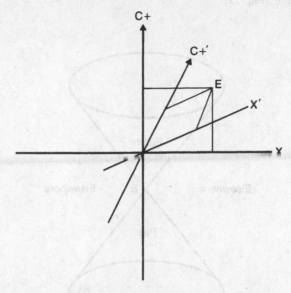

Fig. 11.

of the same event can be designated by the inclined axes x' and ct'.

The equivalence of space and time is expressed mathematically in an equation formulated by H. A. Lorentz. The spatial coordinates x, y, z, and the time coordinate t are related as follows:

$$s^2 = x^2 + y^2 + z^2 - c^2 t^2$$

The mathematician Hermann Minkowski showed that the Lorentz equation can be regarded as rotations of a set of four Cartesian axes in a four-dimensional space-time. The cone defined by $s^2 = 0$ divides the directions of space-time into three classes. The directions which correspond to the three classes are: (i) Space-like, where $s^2 = 0$; (ii) Future time-like where $s^2 < 0$; and (iii) Past time-like, where $s^2 > 0$. Every frame of reference or point in space can be thought of in terms of a light cone with the spatial dimensions horizontal and the time dimension vertical. The double cone is generated by signals travelling into and out of E with the speed of light. Space-time is divided into three regions, future, past, and a region that literally lies beyond space-time which we will label 'elsewhere' (Fig. 12).

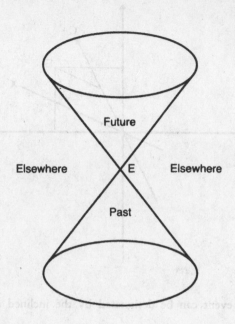

Fig. 12.

In his noted work *Space-Time-Matter*, Hermann Weyl wrote, 'Every world-point is the origin of the double-cone of the active future and the passive past. Whereas in the special theory of relativity these two portions are separated by an intervening region, it is certainly possible in the present case for the cone of the active future to overlap with that of the passive past; so that, in principle, it is possible to experience events now that will in part be an effect of my future resolves and actions. Moreover, it is not impossible for a world-line (in particular, that of my body), although it has a time-like direction at every point, to return to the neighbourhood of a point which it has already once passed through. The result would be a spectral image of the world more fearful than anything the weird fantasy of E. T. A. Hoffman has ever conjured up.'(78)

In units of space and time common to our everyday experience the value of c (the speed of light) is very large. For any observer at point E the light cone would be very flat. In other words, we may think of the 'elsewhere' region is being virtually crushed between the past and the future. (Fig. 13). Olivier Costa de Beauregard

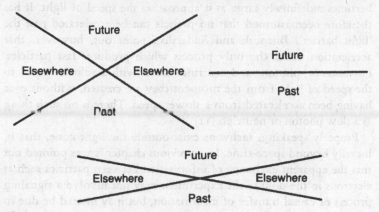

Fig. 13.

reiterates, 'Measured in conventional units, the cone ... is very flat and the *elsewhere* region is very narrow. In the Newtonian limit, in which the finiteness of the speed of light is ignored, this region disappears altogether and space-time consists only of the future and the past, separated by the present instant.'(25)

There are events in our universe, however, where the light cone becomes distorted. For instance, the immense gravitational field created by the collapse of a star creates a state of affairs that Einstein's special theory of relativity is incapable of dealing with. At the end of their lifetime, stars which are greater than 2.5 times the mass of the sun begin to contract and collapse upon themselves. Such stars can be thought of as developing a pucker in the fabric of space-time, a blackhole into which they vanish. The physics of blackholes is poorly understood at present, but there is speculation that black-holes may be 'apertures' to elsewhere, as noted astronomer Carl Sagan calls them.(11)

The implications of something which is 'elsewhere' or literally outside the realm of space-time boggle our imaginations and lead to many startling predictions. It is speculated that there are arbitrarily fast particles or 'tachyons' which, because they move faster than light, exist outside space-time. Olexa-Myron Bilaniuk and E. C. George Sudarshan postulate that Einstein's theory of relativity does not preclude the existence of tachyons; its mathematics even suggests their existence. According to Einstein's theory the mass of a particle becomes indefinitely large as it approaches the speed of light. It has therefore been assumed that no particle can be accelerated past the 'light barrier'. Bilaniuk and Sudarshan point out, however, that acceleration is not the only process which produces fast particles. For instance, photons and neutrinos travel with a velocity equal to the speed of light from the moment they are created, without ever having been accelerated from a slower speed. There is no such thing as a slow photon or neutrino.(11)

Properly speaking, tachyons exist outside the light cone, that is, literally beyond space-time. In a previous chapter it was pointed out that the apparent exchange of information between particles such as electrons in the double-slit experiment may not involve a signaling process or causal transfer of information, but may instead be due to the quantum potential, a hypothetical field that interconnects all the particles so that they function as an indivisible and unbroken whole. If such a field exists, it too transcends space and time as we know them. When a given group of electrons is beamed through the slit, their statistical distribution (which is governed by the collective of electrons) is apparently determined without any reference to time; that is, the statistical distribution is not determined before, during, or after the experiment. The web of interconnectedness that allows the electrons to function as an unbroken whole exists literally beyond time.

For these reasons Jack Sarfatti and Fred Wolf suggest that this information exchange may be tachyonic. In this interpretation a view of the universe, postulated by Feynman and inspired by Everett and Wheeler, plays an important part. Feynman suggests that the continuous space-time world line of an electron (or for that matter, of the entire universe) be combined with the quantum potential. Sarfatti believes that each continuous world line or space-time history is actually just a probability. All possible histories of the universe occur and interfere with each other. The overlap or constructive interference of these 'interpenetrating universes' is the universe we perceive during normal states of consciousness.(67)

Feynman further suggests that the space-time path interpretation of the quantum principle shows that an electron can be scattered *backwards* in time by an electromagnetic vacuum fluctuation. (See Fig. 14, given by Sarfatti.)

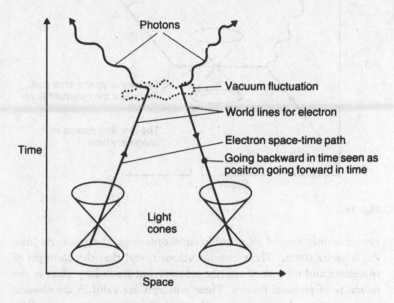

Fig. 14.

An electron moving backwards in time would be detected as a positron of opposite charge, but having the same mass moving forward in time. The new interpretation added by Wolf and Sarfatti is that the electron is also scattered 'outside of its light cone' into a tachyonic world line in which the electron velocity v is greater than the speed of light (Fig. 15).

As Sarfatti states, 'The tachyonic segment of the electron's world line is detected as an *instantaneous discrete quantum jump* which can take zero *real* coordinate time as measured in the laboratory.'(67)

Adolf Grünbaum questions why our time ordering of events is sacrosanct if such faster than light propagations exist.(39) Bilaniuk and Sudarshan reply that they are not sacrosanct. They propose the same tachyonic background utilized by Sarfatti and Wolf – tantamount to an ether reference frame in which 'our' time ordering of

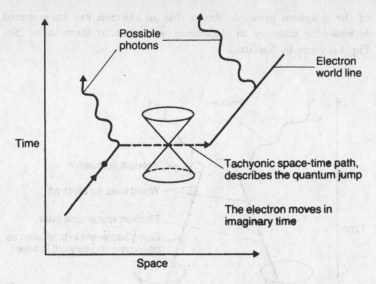

Fig. 15.

events is only one of all possible time orderings of events. As John A. Wheeler states, 'These considerations reveal that the concepts of spacetime and time itself are not primary but secondary ideas in the structure of physical theory. These concepts are valid in the classical approximation. However, they have neither meaning nor application under circumstances when quantum-geometro-dynamical effects become important. Then one has to forgo that view of nature in which every event, past, present, or future, occupies its pre-ordained position in a grand catalog called "spacetime". There is no spacetime, there is no time, there is no before, there is no after. The question what happens "next" is without meaning.'(79)

Once again we are faced with the existential physics of Finkelstein. On the level of everyday life we can function efficaciously according to the absolute of time. But the compass of the physicist again points us towards a direction that our minds have difficulty in conceiving. Phenomena such as the quantum potential and the proposed existence of the tachyon suggest that at least portions of our universe (and – if the bootstrap view of consciousness and reality is correct – portions of our consciousness) exist outside time. The existence of an 'elsewhere' region beyond the light cone and beyond space-time has long been suggested by mystics. Now it is being suggested by physicists. This has awesome implications for our view of the shape of time.

CHAPTER FIVE

The Shape of Time

We must now address ourselves to the problem raised by the necessity for qualifying 'sequence' as nothing more than an 'apparent sequence'. As was suggested earlier, past and future are purely subjective operations and have no objective existence in reality. (The question remains, of course, whether anything exists 'objectively' in reality.) Reality knows only the single frame of the moment of being.

KEITH FLOYD, *'Of Time and the Mind'*

Evidently when Egyptian travelers first encountered the Euphrates they also encountered an unusual difficulty. In the Egyptian language the terms 'north' and 'downstream' are synonymous, the pre-eminence of the Nile (which flows from south to north) made any separation of the two ideas unnecessary. The Egyptian language did not prepare the Egyptians for the Euphrates, which flows from north to south.(26)

Of the many paradoxes presented if a universal tachyonic background is discovered, none are more awesome than the phenomena of retrocausality, or the possibility that an effect may temporally precede its cause. As it stands now, the existence of retrocausality is only a hypothesis. Should it be confirmed, the existence of retro-causal phenomena would be roughly akin to discovering that footprints seen yesterday on a beach were made today. It is like a backwards movie – the footprints come before the walker. In considering retrocausality we find ourselves in the predicament of the Egyptians. Our language cannot easily describe a world in which effect may precede its cause.

In *The Shape of Time*, Yale Art Historian George Kubler examines our concepts of time. As Kubler conceives it, the flow of time can be viewed as assuming the shape of fibrous bundles. Each fiber corresponds to a need in a particular theater of action and the fiber lengths vary according to the duration of each need and the solution to its problems. The cultural bundles of time are therefore composed of variegated fibrous lengths of happening, long and short. Kubler sees them as juxtaposed largely by chance and rarely by conscious forethought or planning.

In light of the findings of quantum theory we have seen that the shape of time proposed by John A. Wheeler is identical to the beautiful imagery presented by Jorge Luis Borges in 'The Garden of the Forking Paths'. Because of the indeterministic nature of *all* events we may view the future world line of our universe as just such a garden. Joseph Gerver points out a possible problem. He asserts that the many worlds hypothesis is true only if one considers universes that split off as one goes forward in time: 'However, if it is possible for the universe to split into two slightly different realities by a quantum-mechanical event, then surely it is equally possible for two slightly different universes to become identical in the same manner. Thus one should also see worlds branching off as one goes backwards in time (indeed, this conclusion is inescapable because of the time symmetry of Schrödinger's equation) . . . we can no longer say that we live in a "normal" or "typical" universe. For if we look at all possible branches going backwards in time, we discover that they look exactly like those going forward in time. That is to say, some of the branches look like the past that we remember, but the overwhelming majority look, more or less, like backwards movies of the future.'(31)

In a rebuttal Bryce DeWitt replies, 'I do not agree with Gerver . . . in his view of the past history of the universal wave function (not to be confused with our own past history, which involves only one branch). The overwhelming majority of past branches would look like 'backwards movies of the future' only if the present state of the universe were the result of a fluctuation from a state of equilibrium in an infinitely old universe.' He adds in closing, '. . . despite the time-reversal invariance, on the quantum level, of the grand Schrödinger equation (neglecting weak interaction) there is no *a priori* reason why the wave function itself should possess a moment of time symmetry.'(31)

In its essence DeWitt's remark is true: there is no reason to suspect that the world line of our universe should possess a moment of time symmetry – a sort of cosmic pivot or moment in history from which we would be able to perceive the entire past as a mirror reflection of the future. However, his statement that our past history involves only one branch reveals a continued prejudice – we conceive of time as having a definite shape in which our history is frozen. Therefore, the only relationship that exists between natural time, the objective time which is the unyielding matrix of our past creations, and subjective time is that of dispassionate observer. We look at objective time and the history of things as we look at a butterfly embedded in glass. We are removed. We feel that we do not participate with the past and this is rightly suspect.

On the quantum level the Schrödinger equation becomes symmetrical in time. That is, it doesn't matter which way the arrow of time is pointing, the equation remains equally as valid for events running forward or backwards through time. It is roughly analogous to a movie of two billiard balls striking each other. In a movie of a black ball rolling across the felt and striking a white ball one can assume two possibilities: (i) The movie is running forward and the black ball struck the white ball; (ii) the movie is running backwards and the white ball actually struck the black ball. In this situation time can be considered 'isotropic': the same laws of physics apply regardless of the direction of measurement.

The same is true of the Schrödinger equation. For any given moment on the world line of a given particle the Schrödinger equation will predict an indefinite number of possibilities extending into the future and an indefinite number of possibilities extending into the past. Depicted in configuration space it appears that the particle's future and past look like two gardens of the forking paths branching out in opposite directions (Fig. 16).

Of course, in reality the particle will be perceived as having only one branch in both the past and the future, just as the experiment with Schrödinger's cat will have only one outcome. But this does not negate the fact that the equation predicts both an indefinite number of futures and an indefinite number of pasts. It has been suggested, for instance, that the universe may contain only one electron! – and that the indistinguishability of all elementary particles of one kind can only be understood in terms of the multi-universe layer picture of space-time. The single electron in our universe is scattered backwards in time as described, but in addition it can actually leave the particular universe layer in which it finds itself and slip into the singular regions beyond space-time. If it should

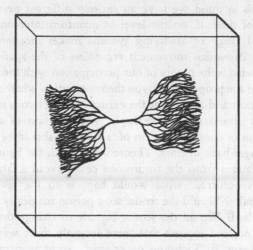

Fig. 16.

return from the absolute 'elsewhere' into its original universe layer or space-time sheet it could confront itself and give the appearance that there are two identical electrons rather than just the original one. The process can repeat itself indefinitely, which accounts for the incredible multitude of identical particles in our universe!(68)

But, of course, the same double garden of the forking paths is true for the world line of our entire universe. Properly speaking, from any (and every) given moment in time the past history of our earth should form a garden of the forking paths. DeWitt states that our past consists of only one branch, but offers no reason why we should accept the shape of time predicted by the Schrödinger equation for the future, but not for the past. Obviously, the answer is because our history books only record one past.

Or do they?

Perhaps there is something to the proposal that the regions of constructive interference of the many 'interpenetrating universes' give us the most probable, classical history of the universe. In other words, perhaps our minds see a consensus reality in such events as Schrödinger's cat because our prejudices are so firmly convinced there is one reality. If the human mind has the ability to edit out one of Schrödinger's cats it would surely have the ability to edit out pasts which make no sense.

With this in mind we have an entirely different perspective on the shape of time. If, on the level of quantum interconnectedness, the general range of all living systems makes incoherent contributions to Brownian movement regardless of the spatio-temporal locations, what is the nature of our participation with the past?

Critics of the proposed tachyon theory question whether retrocausality can exist and argue that the existence of tachyons may present unsolvable paradoxes. For example, with the aid of a proposed tachyon gun or controlled beam of tachyons it should be possible to send messages back in time. Theoretically, with the benefit of such communications from the future one could avoid a fatal accident. But if this occurred, what would happen to the reality of the message sender? Would the sender see a person magically resurrected from the dead? Would the sender be able to change the past? The critics say no, and suggest that there are only three ways to avoid such paradoxes: (i) Tachyons never exist except as virtual particles (which is akin to saying we have a name for such particles, but the particles don't exist); (ii) the universe is constructed so carefully that whenever information is sent into the past it is wiped from the receiver's memory before paradoxes can occur; (iii) emission and reception of tachyons can occur only between members of a limited class of observers possessing velocities relative to some preferred inertial frame (thus never allowing information to spill outside the interim of the present).(10)

The argument against the critics is that they are less concerned with how the universe may be, and more with how they feel it should be. If the major argument against the existence of tachyons is simply that 'tachyons are not logical', we are limiting ourselves considerably. Some physicists feel that no precepts of logic would be violated if tachyons exist and the temporal order of cause and effect is sometimes reversed. In fact, such a reversal in the temporal order of cause and effect may answer more questions than it raises.(10)

In his novel *Childhood's End*, Arthur C. Clarke described humankind's first contact with extraterrestrial life. In the story a fleet of immense ships hover over every major city of the earth for several generations. The Overlords, as they call themselves, are in constant radio contact with the earth, but, oddly, avoid revealing their physical appearance. After many years the Overlords come down from their massive ships. The reason for their seclusion becomes obvious: the Overlords are beings from our mythology, exact likenesses of the devil. When asked if they have visited the earth before and if the legends of the devil are some sort of memory of this visit, they reply, 'It was not precisely a memory. You have already had proof that time is more complex than your science ever imagined. For that memory was not of the past, but of the *future* –'(24)

Interestingly, we can find similar occurrences in history. The Aztecs long possessed legends that the white gods had visited their land. Their mythic leader Quetzalcoatl departed – so the legends say – but promised that he would return on the day of Ce-acatl, in the year of 1 reed. In 1519 Cortes landed at what is now Veracruz and fulfilled the prophecy with unnerving accuracy. Similarly, among the Mayans the legendary white god Kulkulcan left a prophecy to return in the eighth year of Katun 13 Ahau, and in that year (1527) Montejo landed on the east coast of the Yucatán peninsula to begin the slaughter of the Indians. After this, it may be less surprising to learn that in June 1971, when Manuel Elizalde, Jr, led an expedition on Mindanao, in the Philippines, to discover the totally isolated Stone Age tribe of the Tasaday, they were expecting his arrival. Their legends promised that the white gods would someday return.(73)

We may hypothesize that if retrocausality is a fact, perhaps the Spaniards were the *cause* of all the legends that preceded them. Indeed, the subconscious mental functioning of the Spaniards might have participated with the past. The particular bundle of time in which the Aztecs and Mayans possessed such prophetic legends was only one of all possible 'pasts' in the garden of the forking paths.

In *Cultures Beyond the Earth*, Roger W. Wescott formulates what he calls an 'extraterrestrial anthropology' and warns that we should be aware of insights such as Clarke's concerning the nature of time for the universe at large. He proposes that rather than existing in space-time in the conventional sense of the terms, our earth might exist in 'hyperspace' or a space with four or more directions and 'hypertime' or a time which permits events and processes to occur in other than an irreversibly linear and unidirectional manner.

On such a 'hyperhistorical' sphere (or historical 'hypersphere') miraculous beings from our religion and folklore might become explicable intrusions rather than miraculous occurrences. Wescott states, 'From this point of view, extraterrestrial anthropology comes to mean not only interplanetary anthropology, but also the anthropology of our own planet with additional spatial or temporal dimensions added to it – what might be called hyperplanetary anthropology. For such anthropology, the chief supporting disciplines would not be astronomy and astronautics, but as regards its historical aspect, mythology and folklore, and as regards its synchronic aspect, the emerging field of investigation that I call "anomalistics", the systematic study of anomalies.'(55)

Carl G. Jung's postulated archetypes might also be explained by retrocausality. In 1919 Jung theorized that certain 'archaic remnants' or 'primordial images' could be produced by the human psyche. He called these images archetypes and suggested that they were common images that could be found in disparate sources from dreams to ancient myths, religious visions, and fairy tales. In reference to archetypes, Jung states, 'They are without known origin; and they reproduce themselves in any time or in any part of the world – even where transmission by direct descent or "cross fertilization" through migration must be ruled out.'(47)

Jung provided many examples of archetypal information, but throughout his life he continually changed his mind on how to explain such phenomena. At first he believed that there was a 'collective unconscious' or portion of our psyche that was common. This might explain the universality of such images. In 1946 he changed his mind and stated, 'Archetypes . . . have a nature that cannot with certainty be designated as psychic,' i.e., coming from the psyche. Finally, he suggested that archetypes might be 'psychoid' or 'quasi-psychic'. That is, they might be something as much psychic as nonpsychic.(41) In a final change of position, Jung warned, 'My critics have incorrectly assumed that I am dealing with "inherited representations", and on that ground they have dismissed the idea of the archetype as mere superstition. They have failed to take into account the fact that if archetypes were representations that originated in our consciousness (or were acquired by consciousness), we should surely understand them, and not be bewildered and astonished when they present themselves to our consciousness.'(47)

A strange door is opened if the past must be seen as the new physics suggest. Indeed, we cannot even think of ourselves as having a past if our apparent history is just a region of constructive interference of the many interpenetrating universes. If conscious systems affect the Brownian motion of particles regardless of their spatio-temporal location, what limits should we place on our ability to interact with the past? At this very moment our minds might be buffeting around the smoke from Freud's pipe or causing subtle movements in the water beneath some Phoenician ship. And if we are to allow ourselves this much speculation it is not too much to imagine that we might also alter a few words in some old book, or create a painting, or shift from one 'past' to another 'past' as effort-lessly as a phonograph needle switches grooves!

That same portion of the unconscious which enables the high priest of Tamil to walk the coals unharmed may be the elusive misnomer, the collective unconscious, that Jung searched for. In this case, what we are perceiving as archetypes may be images created randomly by the mind and then scattered backwards through time along the garden of the forking paths. The archetype may indeed be psychoid, appearing to come from some collective unconscious, but actually being created in the appropriate consciousness in the appropriate 'pasts'. In this sense the collective unconscious is the reality-structurer.

One is reminded of a quote from Thomas Mann: 'As in a dream it is *our own will* that unconsciously appears as inexorable objective destiny, everything in it proceeding out of ourselves and each of us being the secret theatre manager of our own dreams with us all, *our fate may be the product of our inmost selves*, or our wills, and we are actually *bringing about* what seems to be happening to us.' If our consciousness affects the past, then surely it can affect the future as well. Our very existence in what appears to be a rigidly structured universe may turn out to be more plastic than these few speculations suggest.

The quantum physicists have given us new possible ways of viewing space-time and the history of things. If none of what they say is true, it is an enjoyable fantasy. If their speculations do prove to be true, we must accustom ourselves to an entirely different shape of time: one in which a synthesis of personal and objective time melts our frozen histories. Perhaps the butterflies will be re-leased from the glass of time and we will find ourselves on the shore of a very strange Euphrates indeed.

MYSTICISM AND THE NEW PHYSICS

The realization of the Vedic rishis has become a collective realization; the Supermind has entered the earth-consciousness, descended right down into the physical subconscient, at the frontiers of Matter; there remains but one bridge to cross for the final linking up. *A new world is born. At present we are right in the midst of a transitional period in which the two are mingled: the old world persists, yet all-powerful, continuing to dominate the ordinary consciousness, and the new one slips in quietly, yet very shy, unobserved to the extent that externally it changes little for the moment . . . And yet it works, it grows, till one day it will be sufficiently strong to impose itself visibly.*

SATPREM, *Sri Aurobindo or the Adventure of Consciousness*

CHAPTER SIX

Tantra and Quantum Theory

The generous notions about human understanding ... which are illustrated
by discoveries in atomic physics are not in the nature of things wholly
unfamiliar, wholly unheard of, or new. Even in our own culture they have
a history, and in Buddhist and Hindu thought a more considerable and
central place. What we shall find is an exemplification, an encouragement,
and a refinement of old wisdom.

JULIUS ROBERT OPPENHEIMER, *Science and the Common Understanding*

The concepts presented in the new physics may at first seem strange to
the Western mind. We are not used to interpreting Schrödinger's cats
and the complementarity of nuclear particles. Our minds boggle at
the prospect of curved spaces and regions that literally lie beyond both
space and time. It is little wonder that the new physics ebbs slowly at
the pillars of Newtonian physics. But even more striking than all of
these strange and alien concepts is the fact that they are by no means
new. As Oppenheimer suggests, neither are they wholly unfamiliar.
Indeed, anyone combing the pages of Indian philosophies such as
Tantra is impressed by the unavoidable conclusion that these concepts
have been known for centuries. The purpose of this chapter is to
examine a few of the similarities between Tantra – a Hindu philosophy
believed to have originated around the sixth or seventh century of this
era – and some of the concepts presented in the new physics.

There are many parallel concepts between the ancient philosophies
of the East and the emerging philosophies of the West. Certain
concepts are so similar that it becomes impossible to discern whether
some statements were made by the mystic or the physicist. Esalen
Institute Psychologist Lawrence LeShan gives an example of such an
indistinguishable statement: 'The absolute (is) ... everything that
exists ... this absolute has become the universe ... (as we perceive
it) by coming through time, space and causation. This is the central
idea of (Minkowski) (Advaita) ... Time, space and causation are

like the glass through which the absolute is seen and when it is seen . . . it appears as the universe. Now we at once gather from this that in the universe there is neither time, space nor causation . . . what we may call causation begins, after, if we may be permitted to say so, the degeneration of the absolute into the phenomenal and not before.'(53)

The remark was originally made by mystic S. Vivekananda in *Jnana Yoga*, but the fact that the names of the mathematician who first theorized that space and time are a continuum, Hermann Minkowski, and the greatest of the historical Brahmin sages, Advaita, are interchangeable, demonstrates once again the confluence of mysticism and the new physics.

Vivekananda further expresses a view that has become the backbone of quantum theory: there is no such thing as strict causality. As he states, 'A stone falls and we ask why. This question is possible only on the supposition that nothing happens without a cause. I request you to make this very clear in your minds, for, whenever we ask why anything happens, we are taking for granted that everything that happened must have a why, that is to say, it must have been preceded by something else which acted as the cause. This precedence in succession is what we call the law of causation.'

The Western notion that all actions must have a cause has been one of the major obstacles in our understanding the indeterministic nature of atomic systems. We may suspect that the mystical philosophies hold much more information to aid us in the startling world view slowly being presented by the new physics.

The similarities between Tantra and quantum theory are most enlightening. In fact, divorced from its religious terminology, Tantra can be viewed as an ancient branch of quantum theory. Not only are philosophical sentiments identical, as in the above comparison between Minkowski and Advaita, but mechanical explanations for the working of the universe appear to display a knowledge about physics that we have only recently begun to rediscover. The following are several concepts that are similar from Tantra and quantum theory.

SUPERSPACE VERSUS THE AKASA

Since the time of the ancient Greeks, Western science has tried to understand matter by dividing and redividing it in attempts to discover its fundamental building blocks. One of the basic conceptual problems in modern science has been to understand the building blocks we've discovered, for instance, the fact that subatomic entities such as electrons and protons display the properties of waves and of particles.

But the puzzle doesn't end here. As particles were discovered to be more and more wavelike, phenomena such as light, which had always been interpreted as a wave, became more and more particle-like. At the end of the nineteenth century the German physicist Max Planck suggested that light was discontinuous and consisted of small energy units called quanta. Aside from giving a name to quantum theory, Planck's quanta described light as being a stream of discrete units rather than a continuous wave.

Einstein brought us closer to figuring out the fundamental building blocks of matter when he discovered that light and matter are ultimately interchangeable. The primordial substance of the universe appears to be these wave/particles and quanta. But wave/particles and quanta don't possess any reality, at least not in the terms we are accustomed to dealing with in classical physics. They are both waves *and* particles, two mutually exclusive types of entities, and this complementarity places them in a category analogous to that of Schrödinger's cat.

As we have seen, if the theories of John A. Wheeler are correct we must forsake the degree of reality with which we insist on endowing the physical system.(43) Indeed, we cannot understand the fundamental building blocks of matter in physical terms at all. There is no ultimate physical substance to matter. The primordial substance of Wheeler's superspace is 'something', which we can best understand as being pure geometry.

The Tantric theory of matter is similar. For example, the Hindu concepts of nada and bindu are identical to the concept of matter being both a wave and a particle. Translated roughly, nada means movement or vibration. When Brahma creates matter, nada is the first produced movement in the ideating cosmic consciousness. Bindu literally means a point. According to Tantra, when matter is viewed as separate from consciousness it can be seen as made up of many bindu and physical objects appear to be extended in space. However, when matter is more accurately perceived as being projected by the consciousness, physical objects no longer possess many three-dimensional points in space. Everything collapses to one dimension (remarkably similar to the one-dimensionality the universe takes on if viewed in terms of Wheeler's quantum interconnectedness) and becomes a single-point bindu – or as André Padoux puts it, 'le point sans dimension'.(59) S. Pratyagatmananda wrote that every object or process has to be studied nadawise and binduwise (as a wave or a particle).(64)

John Woodroffe states, 'The Indian theory here described agrees with . . . Western speculations . . . that what the latter calls scientific or ponderable matter does not permanently exist, but says that there are certain motions or forces (five in number) which produce solid matter, and which are ultimately reducible to ether (Akasa). Akasa, however, and scientific "Ether" are not in all respects the same. The latter is an ultimate substance, not "matter", having vibratory movements and affording the medium for the transmission of light. Akasa is one of the gross forces into which the Primordial Power (Prakrti-Sakti) differentiates itself. Objectively considered it is a vibration in and of the substance of Prakrti of which it is a transformation in which the other forces are observed to be operating.'(89)

Prakrti, or the universe of physical objects, is thus seen as being composed of vibration. In essence, the theory of the akasa is identical with Wheeler's quantum foam. Matter is vibrations in the akasa. Matter is undulations in the quantum foam. As Swami Pratyagatmananda observed 'What is appreciated as a "thing" or "stable object" is a power-posture (e.g., physical mass) relatable to a system of power-process (e.g., physical energy functions).'(64)

Lines of Force and the Hairs of Siva

As has been stated, Wheeler conceives of the quantum foam as composed of interpenetrating wormholes which connect all regions of space. Such a paradigm may present a clearer picture of such everyday phenomena as electricity. When electric lines of force converge upon a region of space they don't just criss-cross, but seem to converge and sink into the fabric of space like threads being pulled through a funnel or down a bathtub drain. Wheeler suggests that they must be passing into a wormhole: '*A classical geometrodynamical electric charge is a set of lines of force trapped in the topology of space.*'(79)

This is very startling, for the Tantras speak of space as being permeated by lines of force known as the 'hairs of Siva'.(89) They speak of the hairs of Siva as being able to cause the fabric of space itself to expand and contract. It is utterly incredible that these ancient texts should speak of space in terms of a substance containing trapped lines of force. It is only in the past century that Western science has been able to comprehend space in non-Euclidean terms and tackle the concept of curvature. Yet centuries ago the Tantric texts referred to expansion and contractions in the Akasa! The mathematical conception behind this is stupendous.

Wheeler postulates that electric charge or the lines in an electric or magnetic field are literally lines of force trapped in the fabric of space-time. Sarfatti suggests that the organization of matter is due to a spectrum of self-organizing fields (similar in properties to Harold Saxton Burr's fields of life, also self-organizing) which organizes matter out of the turbulent sea of the quantum foam. Assuredly, the hairs of Siva which organize matter out of the turbulent sea of the akasa are the same thing.

Miniblackholes and the Point Bindu

As has been previously mentioned, Jack Sarfatti has elaborated upon Wheeler's imagery of the bubbles in the quantum foam and hypothesized that they are actually miniblackholes and miniwhiteholes. These miniblackholes and miniwhiteholes (10^{-33} cm diameter, 10^{-5} gm mass) compose curved empty space. Sarfatti states: 'Matter is nothing but gravitationally trapped light. The ring singularity of a rotating blackhole or whitehole is pictured as a photon (particle of light) or a neutrino (another type of elementary particle) moving in a circle ("chasing its own tail"). This occurs because of the self-gravitation of the photon or neutrino. The circular path moves through two space-time sheets, that is, an ordinary universe of positive mass and a ghost universe of negative mass. The two universes are connected through the ring, which acts as a magic looking glass ... When a particle meets an antiparticle to create pure light, the photons that make up the particle and the antiparticle simply escape their traps. The turbulent sea of space of Wheeler's quantum geometrodynamics is simply the trapping and untrapping of photons and neutrinos in a continual process. On this primordial level it is impossible to differentiate among light, matter, and empty space.'(68)

A wormhole or blackhole of macroscopic dimensions presents many startling implications. Indeed, as Wheeler suggests, the existence of macroscopic blackholes has pretty much become an accepted fact. As astronomer Carl Sagan explains, 'At the end of their lifetimes, stars more than about 2.5 times as massive as our Sun undergo a collapse so powerful that no known forces can stop it. The stars develop a pucker in the fabric of space − a "black hole" − into which they disappear.'(66) Two Princeton University scientists theorized the existence of blackholes in 1971. In 1973 the National Aeronautics and Space Administration released a report by a team of scientists from University College, London, concerning the first known blackhole discovered in the double star system of Cygnus X-1, more than 8000 light-years (47 quadrillion miles) away.

Strictly speaking, macroscopic blackholes are literally holes or puckers in the fabric of space and have no actual three dimensions in our universe. The gravitational field a star 2.5 times as massive as our sun would create if it collapsed into a blackhole is so great that not even light would be able to escape from it. It is possible that other stars could fall victim to the powerful gravitational field and slip into the blackhole. The entire universe could be swallowed up in a blackhole and shrink into a point bindu or a point without dimension (Fig. 17).

Where does the universe go when it collapses? According to the Tantric tradition it is withdrawn into the Sakti which created it. It

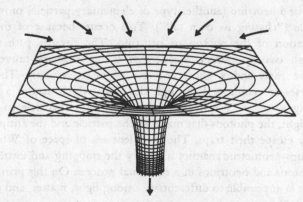

Fig. 17.

collapses into what is known as the Siva bindu, a mathematical point without any magnitude.

Once again a concept in Tantra bears a striking similarity to a concept found in modern physics. The macrophysical blackhole and the Siva bindu are identical. Furthermore, Sarfatti's imagery of a miniblackhole being a photon chasing its own tail parallels a description given by the Tantras. It is conceived that round the Siva bindu there is coiled Sakti. This Sakti may be conceived as a mathematical line which touches the point around which it is coiled on every side. Because the point and the line have no spatial magnitude or dimension they can be conceived of as one and the same, another point bindu. This line is known as the kundalini Sakti because it resembles a kundala ('coil'). The kundalini Sakti is likened to a serpent (Bhujangi) because when it is sleeping it is coiled, and when it is awake it manifests itself in spiraling motion as is witnessed in the revolving orbits of the planets – or as they are known in the Tantras, the 'Brahmanda' or 'eggs of Brahma'.(89)

QUANTUM INNERCONNECTEDNESS AND OMNIJECTIVITY

The most radical assertion made by the new physics is surely that the concept of the 'participator' replaces that of the 'observer'. As has been pointed out, this lack of division between the observer and the observed presents a view of reality that can best be called omnijective. Wheeler's conception of quantum interconnectedness, that every point in space-time is connected via the quantum foam to every other point in space-time, makes our universe into an immense dream space. That is, our perceptions of space and time in a dream exist only to the extent that we conceive them. We may dream of vast spaces, of fields and trees and oceans, but these do not possess any volume. In a dream, as in Wheeler's superspace, all points in dream space and dream time are ultimately connected to all other points via the dreamer.

Sarfatti theorizes that the reality-structurer is based upon the possibility that consciousness is a biogravitational field similar to the gravitational field governing the structure of matter. This is akin to saying that mind and matter are different vibrations or ripples in the

same pond. If this hypothesis is true, we may view the fields which
govern consciousness and those which govern matter as part of a
continuum, a spectrum of fields within fields. On the semantic
level, we may view consciousness and reality as a continuum.
Wheeler's conception of superspace and Sarfatti's own Unified Field
Theory explicitly suggest that the universe is omnijective.

Similarly, the Tantric texts propose that there is no ultimate
division between consciousness and reality. They define three stages
of consciousness one undergoes in approaching this understanding.
The first is a dualistic transformation of consciousness known as
sadasiva or sadakhya-tattva, in which emphasis is laid on the 'This'.
The united consciousness is severed by maya so that the object is
seen apart from the self. The second is ishvara-tattva, in which
emphasis is laid on the 'I'. The third is suddhavidya-tattva, in which
emphasis is laid on both equally and, the Tantras say, illumination
occurs (prakasamatra). The distinction between 'I' and 'This' is no
longer present.

The Tantras assert that the universe may be considered an emana-
tion of the mind. The appearance that it is physical and objective is
'mahamaya', the greatest illusion. But the universe is not a projection
of only one mind. Each of us contributes to the creation of the
projection, say the Tantras.(88)

The views of Wheeler and Sarfatti are again identical with those
of the mystics. Whether it is the participation of those who participate or the general range of all living systems, any interaction between
mind and matter destroys the subjective/objective duality. The distinction between 'I' and 'This' is no longer present and reality must
be viewed as omnijective.

CHAPTER SEVEN

The Interpenetrating Universes

The reality we experience in ordinary states of consciousness is due to the constructive interference of the dynamic phases of 'actions' associated with each of the indefinite number of coexisting universes. I suspect that consciousness may be able to alter the patterns of constructive interference to create separate but equally real realities.

JACK SARFATTI AND BOB TOBEN, *Space-Time and Beyond*

On a rainy day, 13 October 1917, 70,000 people gathered at Cova Da Iria in Fátima, Portugal, to witness a miracle. Six months earlier three children, Lucia Dos Santos and Francis and Jacinta Marto, had first seen the apparition of a lady in a globe of light hovering over a tree. 'Be not frightened,' she had told them. 'I come from heaven.' The holy Virgin of Fátima promised the children that if they would return to the spot at the same time of the month for six consecutive months, on her last appearance she would perform a miraculous feat.

Most of the people in the crowd could not see the Lady of Fátima; only the children shared in that pleasure. But 70,000 onlookers did see something that must shake even the most stalwart sceptic. As they watched, a huge silver disc came through the clouds, rotating rapidly as it drew closer to the throngs of pilgrims. As the object performed aerial tricks it began to change colors and passed through the entire range of the rainbow. Then it passed low over the terrified people and dried their rain-soaked clothing with a wave of miraculous heat. The editor of the Lisbon daily O *Seculo*, Avelino de Almeida, stated, 'The sun "danced", to quote the term used to describe it by the peasants. Most acknowledged that they had seen the trembling and the dance of the sun; others, however, said that they had seen the smiling face of the Blessed Virgin herself, swore that the sun rotated like a firework wheel, that it had come down so low as to scorch the earth with its rays.'(1)

Upon first consideration, the miracle witnessed by 70,000 people at Fátima appears to have two explanations. Either it was a mass hallucination, or the Blessed Virgin created the globe of light which was seen by witnesses as far as forty kilometers away. Let us put aside, for the moment, the notion that it was actually the Blessed Virgin. What are we to make of the term 'mass hallucination'? Such shared visions or *folie à deux* experienced by 70,000 people at Fátima defy explanation within the picture of reality provided to us by classical physics. Only the paradigm of reality created by the new physics offers an explanation. This new paradigm is the subject of this chapter.

In *Flying Saucers*, C. G. Jung recounts how he attended a spiritualistic seance at which four of the five people present vividly saw a globe of light hovering over the medium's abdomen. Jung, the fifth person, saw nothing. Interestingly, he explains that the other four people found it 'absolutely incomprehensible' that he could not see the globe. As Jung points out, what is 'seen with our own eyes' acquires a realness commensurate with our motions of objective reality.(46) What then becomes of the reality of the globe hovering over the medium's abdomen or the miracle of the sun at Fátima? Surely the 70,000 witnesses of Fátima consider their experience real, but within the language and framework of classical physics such visions are implicitly regarded as *not as real as the physical universe*. After the findings of the new physics, this assumption should be regarded as rightly suspect. For too long the term 'mass hallucination' has remained a label and not an explanation.

If we are to truly understand the phenomenon of collective visions we must begin by examining our notions of an objective reality. From the day we are born we are taught that there is a strict commonality to our perceptions. What one person perceives as a tree or a mountain another person must perceive as a tree or a mountain. If there is a disagreement between two perceivers we are explicitly conditioned to suspect that something is wrong. The reason we feel this way is because we believe that there is a physical universe 'out there'. So when the blind men of Cathay stumble across an object and respectively feel a wall, a pillar, a snake, and a vine our minds (indoctrinated by Western structures of thinking) can only comprehend the object as being *one thing* – an elephant. It never crosses our minds that the object might simultaneously be a wall, a pillar, a snake, and a vine. Perceptions must be democratic.

This belief is dramatically demonstrated in experiments undertaken at Harvard concerning the effect of social pressure upon perceptual judgments. When asked to correctly match the length of a line with that of one of three lines presented, participants made the 'wrong' choice less than 1 per cent of the time. However, in a group where the majority was coached beforehand to unanimously choose the 'wrong' line, the decision of the unknowing participants was measurably affected. Under group pressure minority subjects agreed with the majority's 'wrong' judgments 36.8 per cent of the time even when the length of the two allegedly equal lines differed by as much as seven inches. In 'Opinions and Social Pressure', Solomon E. Asch states: 'That we have found the tendency to conformity in our society so strong that reasonably intelligent and well-meaning young people are willing to call white black is a matter of concern.'(2)

Why do we have such a tremendous urge for conformity of our perceptions? Simply because we have taught ourselves to conform. J. R. Smythies points out that the world of the child is quasi-hallucinatory; as they grow up children *learn* to ignore certain aspects of their reality that are considered hallucinatory by the adults around them.(71) In Piaget's *The Child and Reality*, the extent to which perception is learned becomes clearly evident. Time and again Piaget demonstrates that notions of perception being innate or genetic are as yet unproved. The child learns to see geometric forms; the child learns to perceive in three dimensions; the child learns to establish objectal relationships, etc., etc. The ability to perceive may be innate, but it is clear that we learn *what* to perceive.(63)

It should come as little surprise, then, that non-literate societies literally cannot *see* certain types of images such as photographs and films. In a paper presented by Professor John Wilson of the African Institute of London University, Wilson describes how the members of a primitive African village were shown a film intended to teach them methods of sanitation. To Wilson's surprise, not one of the thirty-odd villagers watching the film was able to *see* it. When questioned about what they had seen the villagers were unable to answer except for the curious fact that they had all seen a chicken (which may have possessed some religious significance for them) that had made a momentary appearance in the film. As Wilson puts it, the fowl was the 'one bit of reality for them'.(87)

The eminent cyberneticist Heinz Von Foerster points out that the human mind does not perceive what is 'there', but what it believes should be there. We are able to see because our retinas absorb light from the outside world and convey the signals to the brain. The same is true of all of our sensory receptors. However, our retinas don't see color. They are 'blind', as Von Foerster puts it, to the quality of their stimulation and responsive only to their quantity. He states, 'This should not come as a surprise, for indeed "out there" there is no light and no colour, there are only electromagnetic waves; "out there" there is no sound and no music, there are only periodic variations of the air pressure; "out there" there is no heat and no cold, there are only moving molecules with more or less mean kinetic energy, and so on. Finally, for sure, "out there" there is no pain. Since the physical nature of the stimulus – its *quality* – is not encoded into nervous activity, the fundamental question arises as to how does our brain conjure up the tremendous variety of this colorful world as we experience it any moment while awake, and sometimes in dreams while asleep.'

The answer, of course, is that the brain perceives what it wants to perceive. The truth of this is explicit in the Harvard experiments concerning the lengths of lines, and in the mere fact that children pass through progressive stages of perceptual development – as opposed to being born knowing how and what to perceive. We are not born into the world. We are born into *something that we make into the world*. Or in the words of Von Foerster, 'The Environment as we perceive it is our invention.' Von Foerster, however, does not wish to imply, as he puts it, the 'ridiculous notion of other realities besides "the" only and one reality, our cherished Environment.' To remedy this he postulates that 'cognitive processes do not compute wristwatches or galaxies, but compute at best *descriptions* of such entities'.(37)

Here we arrive at the pivotal issue. It is evident that in the area of perception we have reached a realization analogous to the Heisenberg Uncertainty Principle. We do not observe the physical world. We participate with it. Our senses are not separate from what is 'out there', but are intimately involved in a highly complex feedback process whose final result is to actually *create* what is 'out there'. The pressing question thus becomes: What is 'out there'?

The vast majority of the scientific community, like Von Foerster, maintain that out there is the 'one reality', 'our cherished Environment'. We have grown so accustomed to participating with the universe we have perceptually created that we simply assume that there is an 'out there'. On closer examination the concept of an 'out there' becomes ridiculous. We have no evidence that an 'out there' exists. Indeed, how could we hope to know of the existence of something which lies beyond the senses and by definition cannot be known? The physicists cannot come to our rescue. They have uncovered their own bit of maya. In hoping to find electrons they have found that the consciousness finds what it wants to find. If the field of consciousness is on a continuum with the field of matter-space-time we may expect to find an 'out there' only because we believe it exists. We may also agree with John Lilly when he states, 'In the province of connected minds, what the network believes to be true, either is true or becomes true within certain limits to be found experientially and experimentally.'(54)

We may suspect that the reality of the 'out there' has the same ontological valence as Schrödinger's cat. Everything is grounded on its opposite. If the yes *or* no of Schrödinger's cat is dependent upon which reality the consciousness decides to edit out, the yes *or* no of an 'out there' universe must be assigned to the same category.

In Jung's experience with the collective vision we now find that the globe of light hovering over the medium's abdomen possesses the same realness as the images in the film Wilson showed to the African villagers. We can find a clue to our puzzle in Carlos Castaneda's *Journey to Ixtlan*. Castaneda is an anthropology student chosen by don Juan Matus, a Yaqui Indian sorcerer, to be his apprentice. On 19 August 1961, don Juan informed Carlos that he knew a place in the desert where they could find a spirit. This was the spirit's 'place of power' and don Juan instructed Castaneda to accompany him to the spot. After a lengthy desert search the two of them finally encountered the spirit in the form of a curled-up dog. As Castaneda describes, the creature was large enough to have been a brown calf. However, it was too compact to be a calf and, as don Juan pointed out, its ears were too pointed.

As they watched the animal shivered and Castaneda realized that something was wrong with the beast. As he explains, 'I could not make out its specific features. Don Juan took a couple of cautious steps towards it. I followed him. It was quite dark by then and we had to take more steps in order to keep the animal in view.

' "Watch out," don Juan whispered in my ear. "If it is a dying animal it may leap on us with its last strength." '

As Castaneda watched, the creature writhed iin the last throes of death. It gave several inhuman shrieks as it waved its legs and continued to tremble. As it rolled over upon its back Castaneda saw that the monster was obviously a mammal, but it possessed the beak of a bird. Suddenly, something very strange happened. As Castaneda puts it, 'I wanted don Juan to explain that incredible animal but I could only mumble to him. He was staring at me. I glanced at him and glanced at the animal, and then something in me arranged the world and I knew at once what the animal was. I walked over to it and picked it up. It was a large branch of a bush.'(22)

In our Western way of thinking we may feel a certain sense of relief in Castaneda's discovery – as if we have figured it all out. Don Juan, however, points out something the mystics have known about reality all along. Our relief is a mistake and we have taken a giant step backwards in our search for reality. To the question, was the dying spirit *really* there? Don Juan replies, 'That branch was a real animal and it was alive at the moment the power touched it. Since what kept it alive was power, the trick was, like in *dreaming*, to sustain the sight of it.'(22) If we believe that an 'out there' reality exists, we are probably inclined to react like Castaneda. We arrange the world until we once again perceive our one cherished Environment. We are like Jung and do not see the globe of light. But we should be cautious and not pass judgment on its realness. It is most assuredly like Schrödinger's cat. Its reality and its unreality are solely contingent upon how the collective arranges the world.

What view does this give us of reality? Lawrence LeShan points out that within the framework of our classical view of reality it is clear that phenomena such as Castaneda's dying demon are impossible. Precognition, collective visions, etc., simply cannot exist in the world as we have arranged it and respond to it. 'The problem,' LeShan asserts, 'is that they *do* occur.' As he puts it, 'The evidence, and it is there – hard, scientific, and factual – for anyone who looks at it, is not refutable. We must do something about the paradox.'(53)

To remedy the situation LeShan suggests that there are two realities. The first reality is the reality we have come to know in everyday experience. It consists of solid objects and empty space. It is firmly ensconced in a linear time sequence in which our perceptions are strictly limited by the boundaries of the *present*. And most importantly, in this reality our minds are clearly separate and distinct from the universe about us. To explain the existence of paranormal phenomena LeShan postulates that a second reality *exists*, so to speak. However, this second reality is the reality outside the light cone. It is an 'elsewhere' region and lies beyond our space-time.

The problem with this view is that it is an immense oversimplification. Once again we hold onto our notions of our cherished Environment. We grope around like the blind men of Cathay in search of compact, hierarchical concepts. It is easier to view an object as an elephant than simultaneously as a wall, a pillar, a snake, and a vine. It is easier to believe that there is one reality, our cherished Environment, outside which lies another reality, a paranormal reality, than it is to believe that no such one reality exists. We are led to suspect, as Jack Sarfatti does, that the illusion of the one reality is a result of the constructive interference of all possible realities.

Jerome S. Bruner of Harvard's Center for Cognitive Studies tends not to believe in a world available for 'direct touch'. He postulates that we represent the world to ourselves and then respond to our representations.(19) This is what the mystics have been telling us all along. As Borges suggests, we have dreamed the world. The implications of the new physics destroy our notions of an 'out there' reality. If indeed, the human mind affects which outcome occurs in the experiment with Schrödinger's cat, how could there possibly be one reality? Say, for instance, two individuals who had developed the reality-structuring mechanisms of the human consciousness participated in Schrödinger's experiment. Assuming they have equal control over reality, what would happen if one desired to perceive that Schrödinger's cat survived and the other desired to perceive that Schrödinger's cat died? Surely both would construct the outcome which they desired. They would probably arrive at a situation similar to that encountered by Jung. One would perceive a living cat and one would perceive a dead cat. Suddenly their perceptions would no longer be of an area of constructive interference. They would be perceiving two different, but equally real realities.

In everyday life the illusion of a single reality is the result of the constructive interference of all possible realities. We allow our perceptions to be swayed by the consensus and the consensus determines what phantasmagoria we choose at the arbitrary *one* reality. The paradigm implied by the new physics is that there is no 'out there' reality. Just as the Feynman-Dirac Action Principle proposes that there is no single past history of the universe, in the many worlds hypothesis we may view the history of the universe as having no single present. Thus we have our Schrödinger's cats and our Fátimas.

In the paradigm of the new physics we have dreamed the world. We have dreamed it as enduring, mysterious, visible, omnipresent in space and stable in time, but we have consented to tenuous and eternal intervals of illogicalness in its architecture that we might know it is false. As Joseph Chilton Pearce observes, 'There is no world "out there" available to dispassionate observation. Objectivity in relation to reality is a naive delusion on our part . . . a universal common knowledge is denied. There appears to be no world-mind from which we may get cues, no secret wavelengths for our perceptors.'(61)

Again, the mystics have been telling us this all along. As Castaneda points out in *Tales of Power*, don Juan 'had already made the point that there was no world at large but only a description of the world which we had learned to visualize and take for granted.' According to don Juan's cosmology reality has two aspects, the *tonal* and the *nagual*. In don Juan's way of thinking the *tonal* is everything. In terms of the blind men of Cathay it is the elephant — the single 'something' or form our minds give to the world. It is the wristwatches and the galaxies — the endless hierarchies of gestalts that the consciousness creates for itself to give us the illusion of a single universe. If the *tonal* is the illusory one reality we may view it as the area of constructive interference.

The *nagual*, however, is a much more difficult concept to come to grips with. In terms of the blind men of Cathay we may view it as that *something* which is simultaneously a wall, a pillar, a snake, and a vine. It is Schrödinger's cat, both alive and dead; an area of destructive interference which the *tonal* of our mind struggles to arrange. That is why Castaneda had so much difficulty sustaining the vision of the dying spirit. As don Juan explains, 'When one is dealing with the *nagual*, one should never look into it directly . . . The only way to look at the *nagual* is as if it were a common affair. One must blink in order to break the fixation. Our eyes are the eyes of the *tonal*, or perhaps it would be more accurate to say that our eyes have been trained by the *tonal*, therefore the *tonal* claims them. One of the sources of your bafflement and discomfort is that your *tonal* doesn't let go of your eyes. The day it does, your *nagual* will have won a great battle. Your obsession or, better yet, everyone's obsession is to arrange the world according to the *tonal*'s rules; so every time we are confronted with the *nagual*, we go out of our way to make our eyes stiff and intransigent.'(22)

Pearce arrives at a similar conclusion: Our world is 'word-built' (or very much a creation of the *tonal*). Our reality is a semantic creation largely constructed by our cultural beliefs. Thus, the *tonal* of Wilson's eyes is not necessarily the *tonal* of the eyes of the African villager. To an incredible extent what we believe to be true becomes true. What we call reality is learned.

We may logically expect that we will encounter the *nagual* or areas of destructive interference only in cases where our beliefs come into conflict with one another. We can see tables and chairs and clouds and trees because we all believe in tables and chairs and clouds and trees. But we cannot all see globes of light floating over mediums or dying spirits or the Blessed Virgin – not because they don't exist – but because they don't exist in our *tonal*. In the *nagual* all possible realities coexist in an indefinite number of universes.

The best way of viewing the *nagual* or the paradigm of reality presented in the new physics is to view it as a dream. The assertion that reality is fundamentally dreamlike can be found in many mystical sources. The Tibetan *Madhyamika* maintains that the world should be renounced because it is nonreal as are dreams. A prominent Tibetan sadhana (spiritual discipline) is the Yoga of the Dream State or Mi-lam. In Mi-lam the adept must learn to control every aspect of the dream state. In learning to pass between dream state and waking state at will without any break in the stream of consciousness the Mi-lam adept strives to more completely realize the similar nature of dreams and waking experience.

Control over one's dreams and the realization of the dreamlike nature of reality appears to have one major purpose in most mystical teachings – to loosen the hold the *tonal* has upon the way we construct our environment. As don Juan informs Castaneda, '"Dreaming" entailed cultivating a peculiar control over one's dreams to the extent that the experience undergone in them and those lived in one's waking hours acquired the same pragmatic valence. The sorcerers' allegation was that under the impact of "dreaming" the ordinary criteria to differentiate a dream from reality became inoperative.'(22)

In the paradigm of reality presented by the new physics all categories of real and unreal break down. Just as we can no longer consider Schrödinger's cat as being either alive *or* dead, we cannot consider the objective world as existing *or* not existing. This is exactly what is implied in the Buddhist proverb 'Is God dead?' If you say yes *or* no you lose your Buddha nature. In Buddhism, acquiring a Buddha nature is akin to finally becoming one with the *nagual*. The *tonals* drop from the eyes like ponderous cataracts and suddenly the interpenetrating universes spread out before the consciousness in dazzling splendor.

And what becomes of the 'out there', the one cherished reality? Whatever the consciousness desires. As Pearce puts it, 'Man's mind mirrors a universe which mirrors man's mind.'(61) It was Wheeler's self-reference cosmology that created the dichotomy of mind and universe in the first place – a snake biting its own tail, the participation of those who participate. The world is real only in the sense that it has an objective existence for, and is not a projection of, the individual mind. In a self-reference cosmology, mind and matter coexist. The world of matter is not a projection of the individual mind, but its reality is coordinate with that of the individual mind. In a sense, then, the universe is dreaming itself.

In *Tales of Power* Castaneda sees his double, a spectral image of himself, and asks don Juan whether he was dreaming it or not. Such a question becomes meaningless in a self-reference cosmology. As don Juan replies, '. . . if you had not gotten lost in your indulging, and you could have known then that you yourself are a dream, that your double is dreaming you, in the same fashion that you dreamed him last night.'(22)

The same becomes true of the Lady of Fátima. The realness of the Blessed Virgin, and, indeed, the realness of all the gods and cosmic hierarchies imagined by humankind, acquires the same pragmatic valence as the realness of wristwatches and galaxies. The universe embraces all possibilities because the consciousness can conceive all possibilities. In a self-reference cosmology, as surely as the 70,000 witnesses of Fátima believed and *participated* in the miraculous appear-

ance of the 'dance of the sun' – somewhere there is most assuredly a
Blessed Virgin who believes and *participates* in the 70,000 witnesses.
As in Castaneda's glimpse of his double, the witnesses are dreaming
the vision and the vision is dreaming the witnesses.

We may rightfully suspect that the consciousness is able to alter
the patterns of constructive interference and create separate but
equally real realities. The mystic traditions of all ages speak of being
able to perceive separate realities. In the Sakti Tantras the level of
consciousness required is referred to as turiya, a stage of awareness
in which the dreamlike nature of the world is clearly recognized.
The Sutras refer to this level as samyak-sambodhi. As the Zen
master Hui Hai states, 'Samyak-Sambodhi is the realization of the
identity of form and voidness.'(14)

Or as don Juan puts it, 'The point is to convince the *tonal* that
there are other worlds that can pass in front of the same windows
... So, let your eyes be free; let them be true windows. The eyes
can be the windows to peer into boredom or to peek into that
infinity.'(22)

CHAPTER EIGHT

The Reality-Structurer

Consciousness can act on Matter and transform it. This ultimate conversion of Matter into Consciousness and perhaps one day even of Consciousness into Matter is the aim of the *supramental yoga* of which we shall speak later. But there are many degrees of development of the consciousness-force, from the seeker or aspirant just awakening to the inner urge to the yogi, and even among yogis there are many grades — it is here that the true hierarchy begins.

SATPREM, *Sri Aurobindo or the Adventure of Consciousness*

On the icy slopes of the Himalayas adepts of a hathayogic technique known as *Tum-mo* are said to be able to generate so much body heat that they require little or no clothing. Through various visualizations and breathing exercises the adept imagines a tiny blaze of fire at the base of the spine. With further concentration the adept then causes the fire to extend to the limits of the body and expand to fill the entire universe.

To test the adept's success at *Tum-mo*, the guru may require the adept to sit naked on a mountainside throughout a winter night. During this time the adept must dry sheets dipped in icy water one after another by wrapping them around themselves and subjecting the sheets to the blaze of their psychic heat. Madame Alexandra David-Neel gives accounts of competitions among novice *Tum-mo* adepts who vie with one another to dry as many dripping, icy sheets as possible between evening and sunrise. Accomplished yogins are said to be able to melt a slab of ice several inches thick simply by sitting on it.(28)

To the Western mind such feats as *Tum-mo* are incredible incursions on the reality we have come to intuitively accept. When Borges tells us that we have dreamed the world as omnipresent in space and stable in time, we are willing to accept his vision intellectually. The quantum physicist may concede that there is no physical basis for matter, but rest assured, the search for a physical basis will continue.

After all, a chair is still a chair. The idea that it is ultimately built upon something incomprehensibly wispy and fragile appears to be disproved when we knock our fist against it. The thought that our mind might transcend the solidity of the chair and deal with the super-hologram of swirling wave/particles as if they were a dream image or a phantasmagoria of our imagination is so removed from our experience that it is difficult to fathom it, let alone be amazed by it.

But we have dreamed the world.

And the *Tum-mo* adept sits oblivious to the cold.

However, *Tum-mo* may be only a faint glimmer of the powers available to the human consciousness, if what the physicists are saying is true. For instance, in the super-hologram of reality the consciousness should be able to actually *create* matter. To our Western ways of thinking this is the ultimate miracle – the final proof that the nonphysical, the consciousness, has dominion over the physical world.

Tantric mysticism, both Tibetan and Hindu, has much to say about the structure of matter that parallels the world view of the quantum physicists. Einstein has taught us that matter and energy are convertible: $E = mc^2$, or matter is highly condensed energy. This same view was held by the ancient Tantrists, but with an important extrapolation upon the particular principle. Matter is condensed energy, but is is the condensed energy of *chit*, or consciousness itself. As it is written in the Mundaka Upanishad, 'By energism of Consciousness Brahman is massed; from that Matter is born and from Matter Life and Mind and the worlds.'

Tantra, however, makes a very important point, that reality is ultimately an illusion or maya. In this context we cannot think of matter as either existing or not existing. The consciousness cannot truly create matter – there is no such thing as matter. There is only the constructive interference of the interpenetrating universe.

Don Juan echoes this point when he tells Castaneda that the *tonal* does not create anything. The *tonal* only witnesses. According to don Juan, it is the *nagual* that creates. The *nagual* is the reality that lies beyond our perceptions. It encompasses all possible realities. In Schrödinger's experiment, then, the consciousness does not create a living or a dead cat. It simply decides which universe to witness.

The *nagual* contains both the living and the dead cat and the consciousness merely chooses the *tonal* it wants to perceive. It only witnesses.

'But what's creativity then, don Juan?' Castaneda asked.

'Creativity is this,' don Juan replied. He proceeded to cup his hands and brought them up to the level of Castaneda's eyes. As Castaneda explains, 'It took me an incredibly long time to focus my eyes on his hand. I felt that a transparent membrane was holding my whole body in a fixed position and that I had to break it in order to place my sight on his hand. I struggled until beads of perspiration ran into my eyes. Finally I heard or felt a pop and my eyes and head jerked free. On his right palm there was the most curious rodent I had ever seen.'(22)

Don Juan instructs Castaneda to feel the creature and to Castaneda's surprise he finds that it is a soft, furry, *physical* something. But don Juan soon reveals that the animal is merely a mental construction. As Castaneda states, 'The rodent then started to grow in don Juan's palm. And while my eyes were still filled with tears of laughter, the rodent became so enormous that it disappeared. It literally went out of the frame of my vision.'(22)

In Tibetan mysticism the reality of such mental creations is unquestioned. According to the Tibetan tradition, inasmuch as the mind creates the world of appearances, a master of yoga can create physical objects or *tulpas* by simply developing the powers of the consciousness. As W. Y. Evans-Wentz states, 'The process consists of giving palpable being to a visualization, in very much the same manner as an architect gives concrete expression in three dimensions to his abstract concepts after first having given them expression in the two dimensions of his blue-print.'(33)

In *Magic and Mystery in Tibet*, Alexandra David-Neel gives many accounts of *tulpas* and even recounts an experience of her own in creating thought forms. As she explains, she shut herself in her tent and proceeded to perform the prescribed rites and concentration of thought. After a few months of this ritual she started catching glimpses of a phantom monk, her intended *tulpa*. She states, 'His form grew gradually fixed and life-like looking. He became a kind of guest, living in my apartment. I then broke my seclusion and started for a tour, with my servants and tents. The monk included himself in the party. Though I lived in the open, riding on horseback for miles each day, the illusion persisted. I saw the fat *trapa*, now and then it was not necessary for me to think of him to make him appear. The phantom performed various actions of the kind that are natural to travelers and that I had not commanded. For instance, he walked, stopped, looked around him. The illusion was mostly visual, but sometimes I felt as if a robe was lightly rubbing against me and once a hand seemed to touch my shoulder.'(28)

Madame David-Neel explains that a herdsman even saw the thought form Lama. In time, however, his presence became unwanted and she found to her dismay that it took an equally long time, six months, to dissolve the phantom creation. Once again, however, it is stressed that the reality or unreality of the *tulpa* is not an issue. The basis of Tibetan mysticism is that the worlds and all phenomena we perceive are mirages born from our imagination. The *tulpa* is like Schrödinger's cat. It neither exists nor doesn't exist. Or, as don Juan might say, the *tonal* of Madame David-Neel's phantom monk is only for witnessing.

This is a most important point in Tibetan mysticism. The world is created by the mind. No matter how tightly the *tonals* control our eyes, no matter how firmly we believe in an 'out there' separate and distinct from the consciousness, we must remember this fact. To prove that they firmly realize this, Tibetan disciples of the yoga of visualization subject themselves to a dangerous test – a ritual known as the dance of *chöd*. In the dance of *chöd* the disciple must first have very highly developed powers of visualization. The disciple then finds some secluded place such as a canyon or cemetery and proceeds with the dance. After conjuring up a horde of horrible *tulpa* demons and a *tulpa* or double of the disciple, the ritual begins. The disciple must remain totally tranquil and *will* the demons to attack their *tulpa* double. The disciple must maintain utter composure as the hideous thought forms tear the double limb from limb, disemboweling and devouring it.

If the disciple is firm in the conviction of the dreamlike nature of reality the demons will not be able to harm the disciple in any way. If, however, the disciple's conviction wavers, he or she risks insanity and even death. Confronting demons or *tonals* which lie beyond the constructive interference of our one cherished environment is very dangerous, for one is confronting the *nagual* itself. As don Juan warns, 'No one is capable of surviving a deliberate encounter with the *nagual* without a long training. It takes years to prepare the *tonal* for such an encounter. Ordinarily, if an average man comes face to face with the *nagual* the shock would be so great that he would die. The goal of a warrior's training then is not to teach him to hex or to charm, but to prepare his *tonal* not to crap out. A most difficult accomplishment. A warrior must be taught to be impeccable and thoroughly empty before he could even conceive witnessing the *nagual*.'(22)

Alexandra David-Neel tells of many stories she heard in which Tibetan disciples submitting to the dance of *chöd* were found dead the next morning. She approached an old Tibetan wiseman or gomchen named Kushog Wanchen and asked him if he could shed any light upon these mysterious deaths. To her inquiries the gomchen replied, 'Those who died were killed by fear. Their visions were the creation of their own imagination. He who does not believe in demons would never be killed by them.'

In surprise one of the gomchen's own disciples asked, 'According to that it must also follow that a man who does not believe in the existence of tigers may feel confident that none of them would ever hurt him even if he were confronted by such a beast.'

Kushog Wanchen replied, 'Visualizing mental formations, either voluntarily or not, is a most mysterious process. What becomes of these creations? May it not be that like children born of our flesh, these children of our mind separate their lives from ours, escape our control, and play parts of their own?'(28)

This, of course, is the ultimate message of the Tibetan mystics. If we allow our mental creations to escape our control they acquire a realness equal to the realness of the physical world itself. But if we ignore our dream tigers and *ignore them with the same impeccable attitude that don Juan urges of the warrior*, then, indeed, their reality is transcended. It is like the high priest of Tamil ignoring that fire burns. It is like the *Tum-mo* adept ignoring the cold. The same inner immobility that forces the consciousness to hold tenaciously to the *tonal* can work in reverse.

If what the mystics and the physicists are saying is true, we are upon the threshold of a most remarkable age. The physicists say the role of consciousness has to be reconsidered. The world is omnijective. But most of us are not like the high priest of Tamil or the *Tum-mo* adept. We have little sense, if any, that consciousness is the reality-structurer. We are forced to turn either to the ancient texts or to enlightened masters for the information we seek.

The Tantras say that consciousness and the reality-structurer are synonymous. The reality-structurer or power is represented mythically by Sakti, the female counterpart to Siva. There are many rituals and practices employed to conjure up Sakti, but the basic intent of all of them remains the same. As the Tantras state, there is no difference between Siva as the possessor of power and Sakti as power in itself. The power of consciousness is consciousness, and it is Siva-Sakti that wills the phenomenal universe into existence.

Regardless of the accompanying ritual, the basic intent of Tantric practices for acquiring Sakti is based on the premise that there is no ultimate division between the consciousness and reality. As has been stated, the Tantras define several intellectual stages towards what might be called an omnijective awareness. In the dream state the Upanishads say that the object is manifested in the form of mental states. In the waking state the object is manifested in material states. To acquire power over reality the Upanishads advise that a state of turiya or pure consciousness be attained in which the physical world is no longer seen as being an entity apart from the mind.

In a book on Tantric practices, John Blofeld states, 'The next step is to bring phenomena and mind into a state of perfect unity by discovering the identical nature of waking and dream experience and by seeing that objects and mind, bliss and voidness, the Clear Light and voidness, wisdom and voidness are related to each other like ice to water or like the waves to the sea.'(13)

We may now ask: is the clue to understanding and controlling the reality-structurer an attitude or mental outlook? The answer is yes, but the matter is more complex than this. Very few of us realize the absolute power of the human consciousness. Indeed, that is why we are able to maintain existence in our one cherished Environment. The secret of the reality-structurer *does* lie merely in an attitude, but the *tonal* holds tightly to our eyes. We may rephrase our question and ask: How, then, do we learn to control this attitude?

To answer this we may assume the position that Dr John C. Lilly takes in *The Human Biocomputer*. The cerebral cortex functions as a high-level computer controlling the structurally lower levels of the nervous system. It is a biocomputer. When one uses language or symbols, analyzes, makes metaphors, or, in short, *learns to learn*, one is 'metaprogramming' the human biocomputer.(54) In this light we may view the directions of the ancient texts as metaprograms. When we view ourselves as separate from reality, this view in itself is a metaprogram. If we view dreaming and waking experience as identical, this view is simply another metaprogram. The metaprogram that dreaming and waking experience are identical is preliminary to teaching the biocomputer how to control the reality-structurer.

There are, then, roughly two ways to trigger the reality-structurer. The first can be found in the philosophy behind the visualization rites prescribed by Vajrayana Buddhism. In Vajrayana the yogin undergoes a rigid program of mental control to develop the same powers of visualization demonstrated by Madame David-Neel in her creation of a *tulpa*.

Although the Vajrayana yogins are self-professed atheists, they work their miracles with the aid of various gods and goddesses worshiped in their religious pantheon. Visualization entails choosing an image of one of the gods or goddesses and memorizing its every feature. After months of meditation and prayer the yogin must know every detail of the deity – the hair, the position of the hands, the curve of the smile. Day after day the yogin stares at the image until able to close the eyes and still see the deity in exact detail. Then the yogin continues the rites of visualization until becoming able to imagine seeing the deity as if it were standing before him or her. The yogin must acquire the ability to 'see' the deity as clearly as the furniture or other physical objects in the room. According to Vajrayana, once a yogin has developed the visualization to this extent, just as the herdsman began to see Madame David-Neel's phantom monk, others will also be able to see the deity. As Blofeld states, 'With further practice, it becomes alive like a being seen in a dream. Even that is not enough. As higher states of consciousness supervene, it will be seen to exist in a much more real sense than a person, let alone a dream; moreover, persons like other external objects of perception are of little consequence to the practice, whereas this shining being has power to confer unspeakable bliss and, after union, to remain one with the adept and purify his thoughts and actions.'(12)

The important point is that from the start the Vajrayana yogin remains an atheist, not believing in the entities conjured up; but this does not affect the visualization of the deity. The Vajrayana philosophy recognizes that to trigger the reality-structurer it is important simply to feed the biocomputer the proper symbols. There is an immense power in religion. It is religion that enables the high priest of Tamil to walk the flames. It is religion that enables 70,000 Catholics to alter their reality and perceive the miracle of Fátima in one massive rite of visualization. Religion is a metaprogram, a set of symbols which enables the biocomputer to communicate with the structurally lower levels of the nervous system that control the reality-structurer.

If we cannot acquire the attitude necessary to trigger the reality-structurer, we may, like the Vajrayana yogin, simply choose an arbitrary religion. It is easier to believe in simple cosmic hierarchies than it is to deal with highly abstract notions such as the *nagual* or the void. As Blofeld explains, 'The purpose of visualization is to gain control of the mind, become skilled in creating mental constructions, make contact with power forces (themselves the products of mind) and achieve higher states of consciousness in which the non-existence of own-being and the non-dual nature of reality are transformed from intellectual concepts into experiential consciousness – non-duality is no longer just believed but felt. In short, visualization is a yoga of the mind. It produces quick results by utilizing forces familiar to man only at the deeper levels of consciousness, of which ordinary people rarely become aware except in dreams. These are the forces wherewith mind creates and animates the whole universe; ordinarily they are not ours to command for, until the false ego is negated or unless we employ yogic means to transcend its bounds, our individual minds function, as it were, like small puddles isolated from the great ocean.'(12)

To our Western minds the fact that the Vajrayana yogin confesses to be an atheist and still worships an entire pantheon of deities may seem a bit illogical. But that is why we have so much difficulty in dealing with Schrödinger's cats. Strictly speaking, the Vajrayana yogin is not an atheist as we consider the word, but views the universe as comprising all possibilities. Because all is a projection of the consciousness, the existence or nonexistence of the gods and goddesses is contingent upon one's metaprograms.

Is God dead?

If you say yes *or* no you lose your Buddha nature.

The Vajrayana yogin realizes the first secret in controlling the reality-structurer. Our consciousness is all-powerful. *We* are not all-powerful because we are not in complete control of the consciousness. Because we largely effect what meager control we have over the biocomputer with the aid of symbols, in our evolution toward total control of the consciousness we may continue to feed the biocomputer metaprograms in the form of symbols. Religion is the computer card. Religious belief in a phenomenon is the first method for getting the reality-structurer to create that phenomenon.

As Blofeld points out, the yoga of visualization produces quick results by utilizing forces familiar to us only at the deeper levels of consciousness. Before one can program the structurally lower levels of the nervous system (the levels governing the reality-structurer) one must be able to metaprogram the cerebral cortex with the appropriate set of symbols. This gives us a remarkable new slant on religion. Just as the physicist has found that the nature of matter ultimately defies any single structure or model superimposed upon it, the nature of the universe most assuredly defies any single religious model superimposed upon it. In realizing this, the Vajrayana yogin transcends the use of religion as merely a cosmic hierarchy utilized to explain the universe. Instead, religion becomes our most potent mechanism for controlling the consciousness. Or, as John Lilly states. 'Feelings such as awe and reverence are recognized as biocomputer energy sources rather than as determinants of truth, i.e., of the goodness of fit of models vs. realities.'(54) This is the same point brought out by Joseph Chilton Pearce: our reality is 'word-built' simply because our consciousness *creates* our reality, and consciousness, as we have been taught to *know* it, is primarily experienced linguistically.(61) In short, symbolic systems (religious or otherwise) are the metaprograms which determine how the consciousness constructs the universe.

But there is a second way to trigger the reality-structurer, say the mystics: Bypass the metaprogramming portions of the cerebral cortex and concentrate solely on developing a yoga of control over the central nervous system. According to the Tantras, it is the central nervous system that is the reality-structurer. They maintain that there is an enormous energy locked into the central nervous system. If it is released from the base of the spine it can flow up the spinal column until it reaches the brain. Along the spinal column there are various spinning wheels of psychic energy (chakras) that govern the functions of the body. These are said to be knots by which the soul is tied down to the body. At the base of the spine in the ordinary human this psychic energy is dormant. Mythologically it is represented as a serpent, or kundalini. With the proper meditation techniques the individual can arouse kundalini and move it up progressively through each chakra, untying the knots of the soul until the serpent fire reaches the brain and liberation is achieved.

In his biography of Sri Aurobindo, Satprem observes, 'Generally in the "normal" man these centers are asleep or closed or only allow the smallest little current necessary for his bare existence to filter through; he is really walled up in himself and communicates only indirectly with the outer world, within a very limited circle; in fact, he does not see other men or things, he sees himself in others, himself in things and everywhere; he cannot get out. With yoga the centres open.'(69)

Generally there are two ways that the centers can be opened: the traditional yogic method and the method of yoga as practiced by Sri Aurobindo. With various exercises, the individual can one day feel an energy, much like the intense rush or tingle one feels when listening to a movingly beautiful piece of music, awake at the base of the vertebral column and climb from chakra to chakra right up to the top of the head. The energy will have an undulating movement, much like a serpent, and at each chakra (corresponding anatomically to various nerve bundles of the body) the energy will pierce through and create a spinning sensation (Fig. 18).

In most techniques of yoga the centers are opened progressively from the bottom to the top. In the yoga of Sri Aurobindo, however, the centers are opened from the top to the bottom. In this way, Sri Aurobindo feels that the descending energy opens the centers more gently and slowly. As Satprem writes, 'This process has an advantage if we understand that each centre corresponds to a universal mode of consciousness or energy; if, from the very beginning we open the lower vital and subconscient centres, we risk being now swamped not by our own small personal affairs but by universal torrents of mud; we become automatically connected with the Confusion and the Mud of the world. This is why traditional yogas definitely required the presence of a Master who protects. With the descending

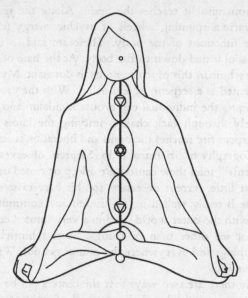

Fig. 18. *(From Sir John Woodroffe,* The Serpent Power, *Dover: New York, 1974)*

Force this danger is avoided and we face the lower centres only after establishing our being solidly in the higher superconscient light. Once in possession of these centres, the seeker begins to know things, the world and himself in their reality, as they are, for he no longer catches external signs, no longer doubtful words, gestures, all that immured dumb show, nor the veiled face of things, but the pure vibration in each thing, each being, at every stage, which nothing can camouflage.'(69)

Gopi Krishna relates that after many years of meditation or metaprogramming he released an energy that exploded in his body with dangerous results. Without the guidance of either a master or any of the ancient texts Gopi Krishna had succeeded in arousing the serpent fire. Unfortunately, because he lacked guidance, he was not able to control the psychic energies and found that his entire body burned with a painful light. It was the very physiological process which went wrong that enabled him to see the whole physiological basis of yogic transformation: a 'biological basis of religion'. In desperation Gopi Krishna wrote to India's most famous saint and sage, Sri Aurobindo himself, and informed him of the painful fire raging through his spine. Sri Aurobindo replied that Gopi Krishna had succeeded in opening the seventh chakra and must find a Tantric yogin to learn its control.(51)

We may be thankful for the self-created dangers such as those Gopi Krishna encountered. We are fanatic in our belief in physical reality, the super-hologram. Indeed, the neurophysiology of the bio-computer is constructed to achieve what might be called a cognitive homeostasis. The eminent cybernetician Heinz Von Foerster points this out: 'The nervous system is organized (or organizes itself) so that it computes a stable reality.'(37)

This is fortunate, for if we should suddenly discover that the mass of the entire space-time construction of the universe is delicately balanced on our minds, we might go quite insane. The human mind sometimes reacts badly to the small shifts in subjective reality that hallucinogens such as LSD create. We cannot experience the *nagual* for very long. Our minds create a 'stable sameness' for themselves and are comforted by this stability.(59) Self-created safety checks in the human nervous system keep us from opening the stargates too quickly.

But open they will. After we have passed through our various dances of *chöd* the convergence of mysticism and the new physics may provide us with the ultimate metaprogram – the metaprogram to create our own realities. It is the nervous system which is the reality-structurer. Each of us participates in the universe and computes the entire space-time construction. The vibrations we perceive as matter, the Brownian movements we unconsciously contribute to – all are creations of the mind. As Sri Aurobindo states, 'The appearance of stability is given by constant repetition and recurrence of the same vibrations and forma-tions.'(8) In his biography of Sri Aurobindo, Satprem continues the master's teachings: '. . . because it is always the same wave-lengths we hook in or rather which hook themselves on to us, according to the laws of our milieu and our education, always the same mental, vital or other vibrations recurring through our centres, which we appropriate automatically, unconsciously, indefinitely; but in reality all is in a state of *constant flux* and all comes to us from a mind vaster than ours, universal; a vital vaster than ours, universal; or from regions lower still, subconscient; or higher, superconscient. Thus this small *frontal being* is surrounded, overshadowed, sustained, traversed and moved by a whole hierarchy of "worlds" . . . or, as Sri Aurobindo says, by a gradation of *planes of consciousness*.'(69)

The teachings of Sri Aurobindo are identical with the intimations of John A. Wheeler. Whether it is the whole hierarchy of worlds or the garden of the forking paths that we contend with, the implications remain the same. The gradation of the planes of consciousness *decide* whether Schrödinger's cat survives the experiment or not. What are we left by all this? As Satprem answers, 'Not much, to tell the truth, or everything, according to the level switched on by our consciousness.'

Humankind is on the threshold of the incredible. The miracle of the high priest of Tamil dwindles in face of the implications of the extent of the reality-structurer. As Pearce observes, 'If a few lone people can reverse causality in isolated cases, what could truly-agreeing people in a mass do with broad statistics?'(61) The answer is, whatever they want to do, to the limits of their creativity. And are there limits to creativity? Yes *and* no. Yes, because the limits themselves are metaprograms and *are* whatever we believe they are up to and including *no limits*.

On the slopes of the Himalayas the *Tum-mo* adept sits and calmly dries his icy sheets. In the words of the Rig Veda (II.24.5): 'Without effort one world moves into the other.'

CHAPTER NINE

The New Cosmology

What then was the commencement of the whole matter? Existence that multiplied itself for sheer delight of being and plunged into numberless trillions of forms so that it might find itself innumerably.

SRI AUROBINDO, *Thoughts and Glimpses*

There is a Hindu myth about the Self of the universe that perceives all of the existence as a form of play. However, since the Self is what there is, and is all that there is, it has no one separate to play with. Thus, according to the Hindu tradition, it plays a cosmic game of hide-and-seek with itself. It assumes a kaleidoscope of faces and façades – a dazzling infinity of masks and forms until it has become the living substance of the entire universe. In this game of hide-and-seek it can experience ten billion lifetimes, see through ten billion eyes, live and die ten billion times. Eventually, however, the Self awakens from its many dreams and remembers its true identity. It is the one and eternal Self of the cosmos. The game begins. The game ends.

By far the most incredible insight we may glean from the convergence of mysticism and the new physics is that in the coming generations our lives may be changed, radically, awesomely. Indeed, if the implications of such a confluence come to pass, life will be transformed into something so *different* that its description is beyond our language. We are on the brink of the miraculous. Our mystics and sages – those guardians of the threshold who have traveled just a bit farther than our reality-structuring will allow – offer us only a vague clue of the infinity of universes that lie beyond.

Most importantly, the new physics is offering us a scientific basis for religion. This is something new in the history of Western civilization, and its impact will certainly be felt in every aspect of our lives. But a word of caution: the religion offered by the new physics is not a religion of values or absolute principles. It offers us no strict delineation of heavens or hells. It is a religion based on the psychology of the human consciousness – indeed, on the psychology of the entire universe as a conscious force acting upon itself. In this new religion we will not find the rules of the game so long sought after by philosophers and theologians. What we will find is a glimpse into ourselves, a bit of cosmic hide-and-seek in which we realize that no rules as such can be found. We make the rules. We play the game. The purpose of this chapter is to examine a few of the rules and then to examine the game itself – the new cosmology suggested by the convergence of mysticism and the new physics.

Anyone delving into mystical avenues of thought will be struck time and again by glimpses of other realities experienced by various people from all times and cultures. From Gilgamesh in the Babylonian epic and John in the Book of Revelation to contemporary people in all walks of life, human beings have experienced worlds that simply don't seem part of our objective universe.

Often these worlds are dreamlike in their splendor and imagery. For instance, in John G. Neihardt's *Black Elk Speaks*, a holy man of the Oglala Sioux gives many descriptions of the visionary realities he experienced. In one he describes, '. . . I was floating head first through the air. My arms were stretched out, and all I saw at first was a single eagle feather right in front of me. Then the feather was a spotted eagle dancing on ahead of me with his wings fluttering, and he was making the shrill whistle that is his. My body did not move at all, but I looked ahead and floated fast toward where I looked.'(58)

Often, as in the case of Black Elk, the visionary reality is interpreted as the *ultimate reality*. More often than not, witnesses to these other worlds perceive them as the rules of the game – the vast and splendid heavenly hierarchies which offer a sort of cosmic truth to some religious system. In *The Tiger's Fangs*, Paul Twitchell describes his experiences of travelling out of his body and into the 'spiritual' realm. In one of his astral travels he describes a visit to the city of Sahasra-dal-Kanswal, as Twitchell puts it, the capital of the spiritual world: 'Everything seemed to be made of soft, white stone that glistened in the reddish light of a sun which was not visible to the eye. There were round, white domes peering over the high walls like oriental temples; and everywhere people walked in joyful steps, heads lifted and eyes sparkling, as though life was completely blissful.

From somewhere came the sounds of beautiful music. Overhead, strange, square-shaped objects flew through the air.'(74)

Because of the incredible sensory impact of heavenly worlds experienced by visionaries such as Black Elk and Paul Twitchell, they are mistakenly interpreted as the one cosmology, the one glimpse of what truly lies beyond space-time. But there is a distinct difference between these other realities and the illusory one physical 'universe' we construct for ourselves. In the physical universe we have rigidly taught ourselves to believe that consciousness is separate and removed from matter-space-time. Thus our eyes are eyes of the *tonal*. But in the visionary realities the consciousness is evidently a different aspect of matter-space-time. Or, we might say, the matter-space-time is more evidently a different aspect of the consciousness. In a reference to the visionary realities or 'Locale II', Robert A. Monroe observes, 'In Locale II, reality is composed of deepest desires and most frantic fears. Thought is action, and no hiding layers of conditioning or inhibition shielded the inner you from others.'(56)

If the human mind has the ability to experience these other realities, where are they? The mystical explanation has always been that consciousness leaves the physical body and travels to them. There are countless occult doctrines of the biological human entity conceived as having one or more spiritual or astral bodies which separate from the physical. This is the position taken by Monroe in referring to the whereabouts of Locale II: 'The most acceptable [explanation] is the wave-vibration concept, which presumes the existence of an infinity of worlds all operating at different frequencies, one of which is this physical world.' As Monroe sees it, this infinity of worlds can occupy the same area occupied by our physical matter world, much as the various wave frequencies in the electromagnetic spectrum can simultaneously occupy the same space, with a minimum of interaction. Only under rare conditions, Monroe asserts, do the many worlds interfere with each other. Thus, as he puts it, 'If we consider this premise, the "where" is answered neatly. "Where" is "here".'(56)

We cannot, however, ignore the psychological aspects of these other worlds. As previously stated, John C. Lilly has described various experiences of these supposed other worlds while isolating himself in a sensory deprivation chamber. Lilly perceives these journeys to other worlds more as journeys deep into one's own mind –

into the metaprograms of the human biocomputer. As he states, 'Possibly one of the safest positions to take with regard to all of these phenomena is that given in this paper, i.e., the formalistic view in which one makes the assumption that the computer itself generates all of the phenomena experienced. This is an acceptable assumption of modern science. This is the so-called *common sense* assumption. This is the assumption acceptable to one's colleagues in science.'(54)

What is the answer? Are these other realities actually places as we conceive of them or do they exist within our heads? In the paradigm of reality offered by both mysticism and the new physics, such a question becomes meaningless. Many mystics caution the student not to think of the so-called 'astral planes' as existing in layers or strata over our own universe. As Swami Panchadasi states, 'The planes do not lie one above the other, in space. They have no spatial distinction or degree.' In a view reminiscent of the peculiar zero-dimensionality transcendent over the three dimensions of John A. Wheeler's superspace, Swami Panchadasi observes, 'They interpenetrate each other in the same point of space. A single point of space may have its manifestations of each and all of the seven planes of being.'(60)

Whether it is the interpenetrating universes of Wheeler's quantum geometrodynamics or the interpenetrating astral planes of Swami Panchadasi's teachings, the message remains the same. In the words of Panchadasi, 'A plane of being is not a place, but a state of being.'(60) The universe itself is not a *place* in the paradigm of the new physics. As don Juan warns, there is no world 'out there', only a description of the world. With the advent of the participator principle the entire matter-space-time continuum of the physical universe becomes merely a *state of being*.

Lawrence LeShan concedes that there are phenomena which do not fall within the classical framework of reality. LeShan proposes that there are two realities – our physical universe, which is subject to distinct laws; and a paranormal reality, in which the normal laws of space and time are violated. LeShan further proposes that we view the entities residing in both types of realities in two distinct ways.

We are 'structural entities', says LeShan. That is, we are entities which have length, breadth, and thickness. As LeShan sees it, structural entities are always subject to the 'normal' laws of space and time. The second type of entities, however, are much different. As LeShan states, 'The second class of things we might call "functional entities". These do not have any length, breadth, or thickness. They cannot be detected by any form of instrumentation although their effects often can be. They are not bound by the "normal" laws of space and time and often can, for example, move faster than light.'(53)

As the perfect example of a functional entity LeShan cites the entities that speak through mediums. The control entities of spiritualist mediums lie in the gray zone between two explanations: (i) beings that are the departed spirits of the dead, and (ii) split personalities created by the consciousness of the medium. Neither explanation fully explains the phenomenon of spirit controls.

Spirit entities do seem to be other consciousnesses. Often they possess information which the medium simply cannot know. Anyone who has ever attended a reputable seance finds too many bits of information coming from the spirit control to be accounted for as merely lucky guesses made by the medium. On the other hand, the psychological element found in spirit controls is undeniable. Few mediums are without controls who fit a sort of romantic myth of what a departed entity must be like. What medium is without an Egyptian princess or Indian guide? When spirit controls follow such simplistic fairy-tale stereotypes, their reality becomes questionable and their psychological and subjective nature becomes explicit.

LeShan suggests that spirit controls are different from either of these two explanations and postulates that the reason we have difficulty in comprehending them is because we try to view them in terms of being structural as opposed to functional: 'The existence of these entities . . . differs considerably from that of the structural entities. They do not have a continuous existence whether or not they are being mentally conceptualized. Indeed, they fit the formulation that Bishop Berkeley attempted to establish – they exist *only* when they are held in the mind; only when being conceptualized, being considered to exist.'(53) Functional entities are thus like a mathematical point. They possess no true reality in space-time, but should be thought of as conceptual aids. The degree of reality of the spirit control is contingent upon the mind of the medium.

The problem with LeShan's point of view is that it is based on a distinction that we simply do not have the right to make. We assume ourselves to be structural entities; we assume the physical universe to be structural. However, in the paradigm of reality presented by the new physics, our concepts of the structural fall to the ground. What structure does an electron possess? None. As Wheeler has pointed out, we cannot think of the particles comprising matter as having locations in space-time. In an existential physics there are no existences – only essences. The particles that comprise matter become little more than a conceptual aid: electrons are assigned to the same category of realness as the points in a line. In the paradigm of the new physics *we are functional entities*. The entire universe acquires the same pragmatic valence as the spirit control and the astral plane. Particles exist when they are conceptualized as existing.

Joseph Chilton Pearce assumes a similar position. As he sees it, spiritualists do 'call up spirits'. They receive 'evidential material' to substantiate the reality of such entities. The instance where medium Arthur Ford called up James Pike's dead son is a prime example. The 'information' is 'there', but as Pearce wisely asks, where is 'there'? He concludes, '*There* is in the act itself, in that particular kind of transaction with reality, that kind of intellectual interaction with possibility. Their very acts, their interactions, may have "produced" the "son" material, much as a Tibetan produces a tulpa figure, or very much as don Juan produces the spirit of the water hole. When that kind of transaction is over, that kind of event might be over. An ordinary traveler passing the water hole perceives no spirits, since he is not interacting with reality in that way.'(62)

Spirit controls lie in the gray zone between 'deceased spirits' and 'split personalities' because they are, in essence, both. The reality-structuring portion of the medium's mind is creating the entity, but simultaneously it is also creating the entire transaction that we refer to as the physical universe. Spirit controls acquire a reality to the extent that they are believed in. The 'there' is, indeed, in the act itself. The 'there' of Black Elk's mystical reality is in the act. The 'there' of Monroe's 'Locale II' is in the act. And the 'there' of the entire matter-space-time continuum *is in the act*. As John A. Wheeler sees it, the vital act is the act of participation – the transaction that we call the interpenetrating universes.

The rules of the game are quite different from *rules* as we know them – as we have been taught to know rules. One must be very careful in this realization. As Pearce sees it, our reality constructs become sort of a cosmic egg which protects us from the arbitrariness of our rules. For instance, when 70,000 witnesses saw the miracle of Fátima they had constructed a definite set of rules for themselves. One of them was that the Blessed Virgin *exists*. And there is nothing wrong with this rule or any other rule. For the witnesses of Fátima the Blessed Virgin became part of their cosmic egg.

The realization that all rules or cosmic eggs are like Schrödinger's cat – that their reality is contingent upon our believing in them, should not necessarily cause us to abandon them. If the 70,000 witnesses of Fátima want to believe in the Blessed Virgin, they should, by all means, continue to believe in the Blessed Virgin. Like the Vajrayana yogins who realize that all gods and goddesses are created by the mind, they should not let this realization force them to discard any cosmic egg. Since the dawn of human consciousness we have taught ourselves to search for the *correct* cosmic egg. The misunderstanding implicit in this search is that there is only one correct cosmic egg.

In the new cosmology, we should learn to accept all cosmic eggs as correct, especially the ones we have chosen for ourselves. When cracks appear in our cosmic eggs our normal reaction is to experience a sort of emotional bankruptcy. This is unnecessary. Cracks in our cosmic eggs are not an indication that they are incorrect. The purpose of the game is not to attain the correct cosmic egg, but simply to be able to pass from cosmic egg to cosmic egg (regardless of how long we choose to remain in any of them) without experiencing emotional bankruptcy. The purpose is not in the attainment, but in the process, the act.

As the Vajrayana texts tell us, no cosmic egg is better than any other. All values are created by the mind. To avoid emotional bankruptcy we must take the position of the Vajrayana yogin and truly neither believe nor disbelieve in any set of rules. This is the pathway to bliss, to Nirvana. In Tantra, to attain Nirvana one must first become aware of the void and nonvoid nature of reality. Just as the quantum physicist must refrain from answering whether Schrödinger's cat is alive or dead, the Buddhist must refrain from making all judgments concerning reality. Thus reality is both void and nonvoid. By recognizing this the Vajrayana Buddhist seeks to control the reality-structuring mechanism of the human consciousness – the mechanism which edits out the infinity of universes we could be experiencing.(12)

When we have attained this control we will truly be on an incredible threshold. As Sri Aurobindo puts it, 'Almighty powers are shut in Nature's cells.'(7) The role of the mystic since time immemorial has been to point this out. One by one, Sri Aurobindo tells us, we may verify that sleep, food, gravity, causes and effects – all the so-called laws of nature – are laws for us only to the extent they are believed. They are like grooves on a record, an infinity of interpenetrating universes. If one changes the consciousness, the groove changes also. As Sri Aurobindo puts it, all our laws are only 'habits'.(69)

What changes might such a drastic transformation in our world view create? In his short story 'Tlön, Uqbar, Orbis Tertius', Jorge Luis Borges describes a mythical society known as Tlön which conceives of the entire universe as being merely a series of mental processes. In Tlön, 'thought' is considered to be a perfect synonym for the cosmos and thus psychology is seen as the major discipline. From this, Borges writes, it would seem possible to deduce that there is no science in Tlön. If the mind creates all laws of nature, how can there be any other study but that of psychology? This, however, is not the case. As Borges explains, 'The paradox . . . is that sciences exist, in countless number. In philosophy, the same thing happens . . . The fact that any philosophical system is bound in advance to be a dialectical game, a *philosophie des Als Ob* [Philosophy of As If], means that systems abound, unbelievable systems, beautifully constructed or else sensational in effect. The metaphysicists of Tlön are not looking for truth, nor even for an approximation of it; they are after a kind of amazement. They consider metaphysics a branch of fantastic literature.'(17)

From the insights of Tlön we may suspect that the same might occur from the convergence of mysticism and the new physics. Our search for a cosmic egg will no longer take us after truth, but after a kind of amazement. The implications are awesome. For instance, if we find that our consciousness acts through some sort of tachyonic background, as Sarfatti suggests, what becomes of the study of history? At present our historians search for the *one true past*. But if we find that the reality of the past is in the same category as Schrödinger's cat and that our minds actually participate with it, there can be no further historical study as we know it. Instead, history may become a sort of game of Tlön. A truly agreeing mass of people could very well get together and decide which history they would like to find and then proceed to make its discovery. As opposed to being a strict science, history might become simply another branch of fantastic literature.

Indeed, it is conceivable that the universe as we know it may cease to exist. In the Introduction a diagram proposed by Carl Jung and Wolfgang Pauli was presented depicting what they felt to be the four major principles governing the structure of the universe. (See Fig. 1 on p. 10.) In Jung and Pauli's *quaternio* most of our present laws of physics are concerned with three of the principles – energy, space-time, and causality.(48) The fourth principle, synchronicity, we may view as representing those acausal phenomena created by the reality-structuring portions of the human consciousness.

At present, the 'universe' we believe in appears to lie closer to being causal than to being synchronous. The synchronous phenomena – the Uri Gellers and the high priests of Tamil – form a small minority. But if the laws of the first three principles prove to be contingent upon the consciousness, we may suspect a slow and continual change of axis from causality to synchronicity. At some point in the distant future causal laws may form the minority. In such a universe consciousness and the other three principles would be seen as synonymous and as Satprem puts it, 'The great Equilibrium is found again.'(69)

In such a 'universe' thought would be the creator of all that is real. The human mind would have the ability to produce *tulpas* at will with the ease of a don Juan or a Tibetan gomchen. Indeed, masses of people, much like the 70,000 witnesses of Fátima, with the proper information could create entirely new 'universes'. As don Juan observes, 'Things are real only after one has learned to agree on their realness.'(22)

If our one cherished Environment is due to the constructive interference of all the interpenetrating universes, what will happen when we cease to agree on the one cherished Environment? Surely various enclaves of individuals (with similar cosmic eggs) will mass together to discover their Locale IIs or their Sahasra-dal-Kanswals.

Indeed, the Blessed Virgins, the UFOs, the angelic visitors, and all the visionary phenomena that have passed before our eyes since the beginning of history may simply be other reality enclaves creating our universe! Our matter-space-time continuum is merely a bubble in the vast sea of the absolute. We may dream ourselves to be omnipresent in space and stable in time, but there are cracks in this cosmic egg. Beyond lies another, more plastic, more luminous – and we are inexorably drawn towards its realization.

This is the secret of all the great mystical teachings. The message has been bandied about until it sounds meaningless and trite – that there are no limits to the power of the consciousness and every human being is the crown of creation. From the new cosmology we may suspect, like Gopi Krishna, that humankind is 'slowly evolving towards a sublime state of consciousness of which fleeting glimpses have been afforded to us by the great seers and mystics of past and present.'(50)

We are temporarily at the causality end of the spectrum and our eyes are still held tightly by the *tonal*. But the super-hologram of matter-space-time is but one of an indefinite number of possibilities. We have dreamt the world and someday it may melt before us and become as hallucinatory as the first perceptions of a child. Momentarily we are locked up in our structure; in the matter of the physical world; as Satprem puts it, '. . . there is the Black Egg which pressures us in on all sides, at every second, and there is but one way of getting out of it, or perhaps two: to sleep (dream, go into an ecstasy, meditate, but all are grades of sleep, more or less noble, more or less conscious, more or less divine) or to die.'(69) And now there is a third way – the way presented by the convergence of mysticism and the new physics. That is, that we simply wake up from our dream.

But whose dream is it? Time and again in the new physics the model of the universe becomes a sort of cosmic giant brain. The holographic properties of consciousness bear an uncanny resemblance to the holographic properties of space-time. As Dr Harold Saxton Burr suggests, in a sort of Unified Field Theory, the various fields within fields within fields that organize the universe appear to have an analog in the biological organization of the L-fields in living cells.(20) David Finkelstein suggests that the elementary particles, building blocks, or 'primitive processes' of matter are assembled into 'chromosome-like code sequences'.(35) Jack Sarfatti points out that the wormhole connections of three-dimensional space connect every part of the universe directly with every other part much like the 'nervous system of a cosmic brain'.(68)

In this cosmic brain the paths of time spread out like the branch-like neurons which interconnect the various portions of the human brain (Fig. 19). The macrocosm is the microcosm. In a universe in

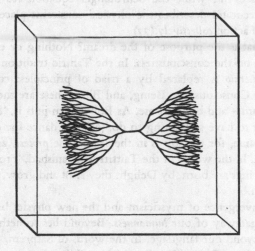

Fig. 19.

which the linear time sequence of the illusory past and the illusory future is due merely to the constructive interference of all possible pasts and futures, the human mind becomes the pivot. It is the filtering mechanism – the synapses in the nervous system of the cosmic brain.

As Sir James Jeans has observed, the universe is a great thought, and we might add, the substance of the great thought is consciousness. As Sri Aurobindo puts it, 'The nescience of Matter is a veiled, an involved or somnambulist consciousness which contains all the latent powers of the Spirit. In every particle, atom, molecule, cell of Matter there lives hidden and works unknown all the omniscience of the Eternal and all the omnipotence of the Infinite.'(5) Evan Harris Walker echoes Aurobindo's sentiments when he notes that the behavior of elementary particles appears to be governed by some sort of conscious force. As Walker sees it, 'Consciousness may be associated with all quantum mechanical processes.' He continues, 'Indeed, since everything that occurs is ultimately the result of one or more quantum mechanical events, the universe is "inhabited" by an almost unlimited number of rather discrete conscious, usually nonthinking entities that are responsible for the detailed working of the universe. These conscious entities determine (or exist concurrently with the determination) *singly* the outcome of each quantum mechanical event, while the Schrödinger equation (to the extent that is is accurate) describes the physical constraint placed on their freedom of action *collectively*.(77)

And what is the purpose of the dream? Nothing or everything, depending on the consciousness. In the Tantric tradition, Jung and Pauli's *quaternio* is replaced by a triad of principles: cit, sat, and ananda, or Consciousness, Being, and Bliss. These are the substance of the dream – and the purpose. As Bob Toben puts it, 'Space-time is here just to have something to do.'(68) We dance the dance. We play the game, for the joy is in the *change*, the *process*, and not the attainment. In the words of the Taittiriya Upanishad, 'From Delight all these beings are born, by Delight they exist and grow, to Delight they return.'

The convergence of mysticism and the new physics has brought us to the gateway of our *humanness*. Beyond lies something that is literally beyond our language. In the words of Satprem. 'We are at the beginning of the "Vast" which will be always vaster. The pioneers of evolution have already found other grades in the Supermind, a new curve is taken in the eternal Becoming. At every conquered height everything changes, it is a reversal of consciousness, a new heaven, a new earth; the physical world itself will change soon before our incredulous eyes. And this is not perhaps the first change in history, how many must there have been before ours? How many even can be *with* us, if only we consent to become conscious?'(69)

There is a Tibetan legend about King Gesar of Ling who was able to create *tulpas* with great ease. He would multiply himself into hundreds of horses, servants, lamas, merchants, phantom tents, and caravans for the simple delight of experiencing all forms. This is the secret of the cosmic game of hide-and-seek. The consciousness behind the universe is dreaming us, ten billion phantasmagoria in a kaleidoscope of thought. And within us we grow increasingly aware of other universes, worlds, and entities which we are dreaming. For the absolute is a dream within a dream within a dream, to which there is no beginning and no end. In the words of Sri Aurobindo, 'But what after all, behind appearance, is this seeming mystery? We can see that it is the Consciousness which had lost itself returning to itself, emerging out of its giant self-forgetfulness, slowly, painfully, as a Life that is would-be sentient, to be more than sentient, to be again divinely self-conscious, free, infinite, immortal.'(6)

This is the secret. Each time the kaleidoscope turns over in our head we see a universe, new and unexpected. We travel through the worlds in search for something that forever lies within us – the 'childlike laughter of the Infinite', the eternal delight of the game player. Our Evolution has not come to an end. It is not an absurd round, but an adventure in Consciousness. The Great Bliss continuously arises. If we simply realize our infinite ability to experience, the thousand-petaled lotus will unfold before our eyes. And *someday*, as Bob Toben so beautifully puts it, 'we won't stop smiling. When we walk, we'll float. And light will pour from our eyes.'(68)

CHAPTER TEN

An Afternote On Language

There is something maddening about answers to metaphysical questions. Anyone who has ever puzzled over a Zen proverb must know this frustration. Time and again the 'answer' seems to be there, but one cannot find it. It is as if it is lost in semantical hocus pocus. The wisdom of the ancient teachings hides behind the cosmically ambiguous, and all that saves us from throwing up our hands is the promise that the 'answer' *is* there, it simply cannot be put into words.

One i tempted to heed proposition 7 of Wittgenstein's *Tractatus Logico-Philosophicus*, 'Of which we cannot speak we have to remain silent.' But we don't remain silent, and we never will. This is not merely a dilemma of the metaphysician, but a criterion of consciousness. The mind *is* an infinite space.

The reason we should not remain silent is simple. We are frustrated by the Zen proverb because it imparts no information. It is like Schrödinger's wave function. Take, for instance, the Zen proverb, 'Is God dead? If you say yes *or* no you lose your Buddha nature.' Such advice imparts no informmation. It is cosmically ambiguous and presents a yes *and* no logic very much like the Everett-Wheeler interpretation of quantum physics. Is Schrödinger's cat alive or dead? If you say yes *or* no you lose your Buddha nature. Our intuitions are assaulted, we receive no information, but silence is not necessarily the answer. The true value of the Zen proverb and the answer to the metaphysical question are not in the information they pretend to deliver, but in the effect they create within the consciousness.

There are two interesting points to be learned from the convergence of mysticism and the new physics. The first is that the ultimate nature of reality transcends language. Time and again the paradoxes met in the new physics reveal this. The complementarity of wave/particles, the yes *and* no logic of the quantum principle, the beyond real and unreal nature of the interpenetrating universes – all transcend the limited dualities of our language.

That is because language is based upon discrimination. This is delightfully illustrated in a little story about Yungchia and Huineng, the Sixth Patriarch in Zen Buddhism. The two are engaged in a little conversation in which Yungchia remarks, 'Birth-and-death is a problem of great moment; all changes ceaselessly.'

'Why not embody impermanence, and so solve it?' Huineng replies.

'To be unborn and deathless is to embody it,' Yungchia continues, 'to be timeless is to solve it.'

'That is so, that is so,' conceded the Patriarch. 'Aren't you in a bit of a hurry to be off?'

Yungchia thought a moment and then retorted, 'Motion has no real existence, so how can there be such a thing as "hurry"?'

'Who knows that motion is unreal?' the Patriarch countered. 'You yourself are discriminating in asking such a question. You have grasped birthlessness splendidly!'

But Yungchia remarked, 'Has the expression "birthlessness" any meaning whatever?'

'If it had no meaning, how could anybody discriminate?' the Patriarch asked.

'Discrimination is meaning!' Yungchia cried.

'Very good indeed!' exclaimed the Patriarch.

When the physicist discovers the indistinguishable nature of electrons, this realization becomes painfully evident. Two electrons can either be referred to as the 'same' or 'different' and neither word imparts any more information about the phenomenon of the electron than the other. Discrimination is meaning. If one cannot discriminate between two electrons there is no meaning in words ascribed to their difference.

We pretend to attach our words to our perceptions. Thus, when we encounter phenomena in which our perceptions cannot discriminate, we feel that our words have lost their alleged meaning. The truth of the matter, of course, is that words do not possess any meaning or information to begin with. All information is in the head. As John Brockman states, 'What is the information from an electric light bulb? No information. What is the information from a book? No information. To speak of a change as giving information implies that there is somewhere a receiver able to react appropriately to the change. Be concerned only with the changes in the operations of the receiver, the brain, in terms of the transactional present. Do not confuse information with signals or the source of signals. The mind of the observer-participant is where the information is constructed, by and through his own programs, his own rules of perception, his own cognitive and logical processes, his own metaprogram of priorities among programs. His own vast internal computer constructs information from signals and stored bits of signals. Information is a process. There are no sources of information; there are no linear movements of information to the brain.'(18)

Thus, when we look again at the Zen proverb we cannot criticize it for not imparting information, because no words impart any information. The purpose of the Zen proverb is to arrest the reader's mind and throw it off of its familiar metaprograms of Western reasoning. Indeed, once we realize this the paradoxes and contradictions of the Zen proverb begin to say a startling and powerful statement on the nature of symbolic thinking. They once again *seem* to impart information. But let us not be fooled. Our mind has flip-flopped those little perceptual gestalts we call words and figured out a way to create the illusion that they possess meaning. But there is no meaning there, only discrimination.

But we are still faced with the first problem, that in both mysticism and the new physics we have reached the limits of our language. As Heisenberg puts it, 'The problems of language here are really serious. We wish to speak in some way about the structure of the atoms ... But we cannot speak about atoms in ordinary language.' The mystics knew this long ago. As Robert Sohl and Audrey Carr state in *Games Zen Masters Play*, 'To confuse the indivisible nature of reality with the differentiations and conceptual pigeonholes of language is the basic ignorance from which Zen seeks to free us. The ultimate answers to existence are not to be found in intellectual concepts and philosophies, however sophisticated, but rather in a level of direct nonconceptual experience which can never be limited to the dualistic nature of language.'(15)

This, indeed, is the problem, and this brings us to the second point. If we have reached the limits of our language, are there modes of thinking beyond language which we can use to discover the alleged 'ultimate answers' to existence? Yes, say the mystics.

John Lilly proposes that we acquired our language making abilities when we developed our cerebral cortex. It is the high-level software computer controlling the structurally lower levels of the nervous system. In *Philosophy in a New Key*, Susanne K. Langer asserts that the entirety of human experience can be viewed from the standpoint of the development of these abilities,(52) which places us in a most peculiar position. We have discovered that our language limits our experience of the 'realities' encountered in both mysticism and the new physics; and yet our linguistic ways of thinking dominate our life. We do not realize that outside the narrow plane of words there may be vast realms of conscious experience which we are denying ourselves. We have been culturally conditioned to *think with words*.

But there are other ways of thinking. As Joseph Chilton Pearce states, '. . . our biological development keeps our options open to the "given state" in spite of our cultural conditioning. Even as we seal ourselves into a word-built world, one function of our intellect breaks that seal and keeps our lines open. These "lines" lie not so much in our head as in what don Juan calls our "body-knowing".'(62)

A similar assertion is made by Karl H. Pribram, who feels that all thinking has, in addition to sign and symbol manipulation, a holographic component. As he sees it, holograms are 'catalysts' of linguistic thought. Though they remain unchanged, according to Pribram, they enter into and facilitate the thought process.(65)

It is as if we are on a long journey of rediscovering the structurally lower portions of the human nervous system. It is the cerebral cortex which gave us our language making abilities, but to experience reality beyond our verbal training we must retreat to the body-knowing – those holographic portions of thought which literally lie beyond words. As in the *Song of Mahamudra*, by the Tibetan Buddhist Tilopa (988–1069):

> Though words are spoken to explain the Void,
> The Void as such can never be expressed.
> Though we say 'the mind is a bright light',
> It is beyond all words and symbols.
> Although the mind is void in essence,
> All things it embraces and contains.(23)

There are many ways to make the retreat back into the nonverbal realm and it is not the purpose of this book to cover them here. All that should be remembered is that the confluence of mysticism and the new physics has pointed out that our plane of linguistic thinking is a mixed blessing. As Langer points out, we measure the history of our experience according to its evolution. But we are drawing close to a time when linguistic thinking will begin to hinder our evolution. We must remember, as the mystics tell us, that the Mahamudra is the great symbol *beyond all words*. Just as we may teach our eyes to let go of the *tonals* we may teach our minds to let go of the universe of words. For it is written, 'Mahamudra is like a mind that clings to nought. Thus practicing, in time you will reach Buddhahood.'

Mysticism and the New Physics Revisited

HOW THIS BOOK CAME TO BE WRITTEN

It is strange writing a new afterword to a book I wrote so long ago, peculiar mainly because it necessitated reading the book and I usually do not read my own books. This is for two reasons. First, by the time I have seen a book through to publication I am usually so saturated with the information it contains that the last thing I want to do is read the material yet again. Second, I am such a perfectionist I am never happy with anything I write. As a consequence, reading my own words after they are published (and can no longer be tinkered with) is always painful for me because all I see are ways that I could have phrased something better, or things I should have added. Rereading *Mysticism and the New Physics* was no different. I cringed often. Such is the fate of a perfectionist.

However, I also found myself strangely entranced because I wrote the book so long ago I had forgotten much of what was in it. Thus it was like reading someone else's book and this was eerie. I kept thinking, *I wrote this?* Occasionally, between cringes, I was even impressed by an idea or turn of phrase here and there. It must be similar to what individuals with multiple personality disorder experience when they encounter a poem or drawing executed by one of their 'other' selves.

The observant reader may notice that this book was first published in 1981 and hence my reference to the great amount of time that has passed since I first wrote the book may seem a bit specious. But the truth is I wrote *Mysticism and the New Physics* almost twenty years ago and therein lies a story. Since it is an interesting story — and writing this new afterword affords me the opportunity to relate it — I'm going to tell it.

As readers of some of my other books may know, my interest in the confluence between science and the mystical comes from the many psychic and paranormal experiences I have had throughout my life. As a small child I spontaneously remembered what appeared to be memories of former existences. These past-life memories were so vivid to me that I refused to call my parents 'mother' and 'father'. I had such a firm sense that I had had other parents in other times and places that I was confused as to why these two kind people were asserting *they* were my parents. I started calling them mother and father only after they explained that my insistence on addressing them by their first names was embarrassing them in front of their friends.

As a child I was also compulsively drawn to things having to do with the Far East, preferred sitting cross-legged on the floor to sitting in chairs, insisted on drinking several cups of strong black tea every day, and spontaneously recited Buddhist aphorisms and prayers. Several of the lifetimes I recalled were of incarnations in various Eastern settings, including India, China, and Tibet, and I believe this is why I have always had a powerful resonance for Eastern philosophies – a resonance that is reflected in the book.

I also think that it is more than coincidence that many of the lifetimes I recall were spent engaged in spiritual development and in pursuit of spiritual answers (as monks, priests, and so on). It is my belief that souls sometimes gravitate towards the same interests over a series of incarnations. Some may spend several lifetimes exploring military existences. Some, exploring different avenues of musical expression. If I am interpreting my apparent memories correctly, my soul has had a tendency to explore spiritual issues over a series of lifetimes. In some of these existences I was also very much concerned with developing various psychic abilities and I believe that is in part why I have had so many psychic and paranormal experiences in this life.

These experiences have manifested in many different ways. For example, throughout my childhood and into my twenties I was the focus of an active poltergeist haunting, a type of haunting that centers around people rather than places and manifests mainly through loud, inexplicable noises and displays of psychokinesis, or mind over matter. My mother tells me that even when I was an infant pots and pans were already jumping off tables by themselves. When I was five-years-old the haunting manifested by occasionally showering the roof of our house with gravel, and later by pelting me with stones and other small objects while I was inside my house. I have written about many of these experiences in greater detail in my book *Beyond the Quantum*.

Many researchers believe that poltergeists are actually unconscious expressions of the latent psychic ability of their human centers and I believe this is true. One reason is that the older I grew, and the more conscious I became of my psychic abilities, the more the poltergeist activity waned. I believe this is because it was indeed an expression of my unconscious; as long as I remained unaware of the psychic currents running through me, my unconscious mind had no choice but to sublimate them and express them only via occasional and seemingly nonsensical displays of psychokinesis. But as I grew more aware of my abilities I had no longer had any need for such an escape valve, so to speak.

The poltergeist is not the only way my psychic abilities manifested. Throughout my life I have also had numerous other brushes with the paranormal, out-of-body experiences, frequent and uncannily accurate premonitions of future events, and even occasional visions and shamanic encounters with otherworldly levels of reality. In fact, such events occur so frequently in my life that as a teenager it came as quite a shock to me to discover that other people weren't having such experiences daily. I was not unlike a child who doesn't realize he needs glasses because he thought it was normal for the world to look fuzzy. Only I thought it was normal to receive occasional visits from spirits and to experience constant evidence that the consciousness can affect and directly interact with physical reality.

I have also always had an overdeveloped sense of wonder, and spent a good deal of my youth staring through microscopes and telescopes, performing chemistry experiments, gazing at insects through magnifying glasses, gathering specimens while wading up to my knees through woodland ponds, and generally reading anything I could get my hands on about science. I was and still am obsessed with understanding the universe in scientific terms.

Oddly, it wasn't until I was nineteen years old and working on a B.A. in graphic arts in college that I realized what a dichotomy I represented. On one hand I believed (and still believe) that science is one of the most powerful tools the human race has invented for exploring the universe. But on the other, I experienced daily a world that science not only did not explain, but for the most part did not even acknowledge exists. This bothered me for I have always believed that the supernatural is just the natural not yet understood. It was naïve of me not to realize that I had a foot in two worlds, but again, I had always viewed the activities of the poltergeist with the same mixture of acceptance and wonder with which I viewed the birth of a school of protozoa out of a seemingly inanimate handful of dried mud, or the movement of blood corpuscles through the delicate vein in a butterfly's wing. It was on the day that this realization struck that I embarked on the path that ultimately led to the writing of *Mysticism and the New Physics*.

I spent the next year reappraising the sticks-and-stones view of the world I had been taught in my high school science classes and the more plastic reality of my day-to-day experience; and formulating a paradigm of reality that acknowledged and incorporated the existence of the psychic and the spiritual. As I did this it became apparent to me that much of what I had been taught was wrong. If, as my experience demonstrated, the mind occasionally was able to access the future via precognitive experiences, then time was not as absolute or inviolable as I had learned. If, as my experience demonstrated, the poltergeist sometimes was able to dematerialize an object and cause it to rematerialize in an entirely different location, then the objective world was not as solid and immutable as it seemed.

In 1974, when I finished this reappraisal, I realized that I believed in a picture of reality that was very different from the one embraced by most people. To begin, in order to preserve the concept of free will and yet still acknowledge that at least some aspects of the future have already coalesced and can be accessed from the present, I concluded that more than one reality must exist. It occurred to me that, in a sense, many parallel universes must exist and the human mind itself must play some role in determining which of these probabilistic futures manifests as real. I also concluded that reality was far less substantive than it appeared and that, again, the human mind played some role in this process, although I suspected this role was deeply unconscious, involving perhaps even deeper strata of the psyche than those involved in autonomic functions such as digestion and the regulation of the glandular system.

I knew that the picture of reality I had arrived at was not unique and had in fact been proposed in slightly different ways by numerous mystical traditions of the past, most notably in Hindu thought with its concept that all reality is an illusion of the mind or *maya*. But I remained deeply troubled by the fact that science seemed so little aware of what I felt was a more accurate way of looking at reality.

One reason I was disturbed was that, although I am a deeply spiritual person, I am not a religious person. I believe that we live in a spiritual universe but feel very strongly that the best way to understand the spiritual aspects of reality is by exploring them scientifically – by forming verifiable theories, or at least educated guesses based on credible evidence, and allowing those theories to fall by the wayside when newer better theories take their place – and most definitely not by trying to freeze our spiritual ideas in a stagnant glacier of religious dogma.

None the less, so radical were the revisions I felt were necessary in order to amend our current picture of reality that I could not help but feel that some hint of them must have already been encountered by science. If this were true, it occurred to me that the branch of science most likely to be addressing these concerns was the one that dealt with the most subtle and fundamental aspects of reality, or quantum physics.

It was a warm August day in 1974 when this thought occurred to me and a few days later I went to the physics library at Michigan State University to determine if my assumption was correct. Once there I employed a technique that I have practiced often and with great success throughout my life. Instead of searching out the answer with the aid of my conscious mind, I began by first relying on my deeper and more intuitive abilities.

To do this I set off aimlessly through the labyrinth of book shelves. As I did so, I did not look at any of the titles but instead waited for a book to 'call' me. Several minutes later and without any conscious intervention on my part, I felt a sudden compulsion to stop. Just as abruptly my hand reached out and grabbed a volume off the shelf, seemingly opening it to a page at random.

Only then did I look at the book. I discovered that I had taken down a bound set of *Physics Today* magazines and had opened it to a September 1970 article titled 'Quantum Mechanics and Reality', written by University of North Carolina physicist Bryce S. DeWitt. In it DeWitt explained that quantum physicists had discovered not only evidence that the existence of reality is dependent on the human mind, but also evidence suggesting that subatomic events split the universe into an incomprehensible number of parallel universes.

Indeed, as I hungrily devoured the article I discovered that quantum physicists had come to many of the same conclusions about reality I had arrived at, only working from a completely different direction and based on an entirely different class of phenomena. Words cannot express the joy, even the familiarity, I experienced at encountering the strange and wonderful ideas offered by quantum physics. Although I was confident in my own thinking I had come to feel quite alone in my strange views of reality. DeWitt's article, and the books its bibliography led me to, not only made me realize that others also had recognized that our current picture of reality was, in a sense, the emperor who has no clothes, but launched me on a passionate study of quantum physics which continues to this day.

As noted in the book, Jungian psychologists call such eerily meaningful coincidences – like my plucking 'at random' the precise book and page I needed to answer my question – synchronicities. Interestingly, our current cultural mind-set tells us that synchronicities are unusual occurrences and yet I cannot begin to relate how many dozens of people have told me identical stories, accounts of how they walked into a bookstore or library and gravitated mysteriously but unerringly toward precisely the book or article they needed to answer some pressing question. It is the process on which the Chinese book of divination, the *I Ching*, is based. As the Bible puts it, 'Ask and ye shall receive.'

Jung himself noted that synchronicities tend to clump around times when an individual is on the verge of some profound breakthrough or transformation, and surely this is true, for my discovery of the DeWitt article was not the only synchronicity I experienced on that watershed day. After leaving the physics library I ran into an acquaintance walking on campus, and accompanying him was a young woman. The acquaintance, a man I knew only slightly and who knew nothing about my interest in the paranormal, introduced me to the young woman, and after we exchanged amenities the woman suddenly gave me an odd stare. 'I have a letter you're supposed to read,' she blurted out cryptically. I asked her what she meant and she apologized for her outburst. 'I don't know,' she explained. 'I know it sounds strange and I don't even know why I'm telling you this. All I know is that I have a letter you're supposed to read.'

There was such a sense of urgency in her voice that we all went back to her apartment. Once there she showed me the letter. It was dated 9 August 1974 and was written by a friend of hers who was a physics graduate student at the University of Ohio. In the letter (I still have a photocopy of it) he explained how excited he was over some of the findings quantum physicists were making regarding reality. In particular he was especially impressed by the work of John Wheeler, the Princeton University physicist responsible for the notion that our minds do not observe reality, but *participate* with it. Thus another link in the chain was forged.

Within days of frenzied reading – at the time Capra had not yet published his now seminal *The Tao of Physics* – it struck me that the ideas inherent in the new physics not only had profound mystical implication, but also deserved a far wider audience than they were receiving. As a result, I set about writing *Mysticism and the New Physics* almost immediately. By the time I finished writing the book Capra had already published *The Tao of Physics* and my literary agent at the time sold *Mysticism and the New Physics* to Bantam in 1976, shortly after Bantam picked up paperback rights to Capra's work. For reasons known only to themselves, Bantam waited five years before publishing *Mysticism and the New Physics*. They were contractually obligated to publish the book within two years, but after that time elapsed my then literary agent informed me that he did not feel he could re-sell the work and advised me to stay with Bantam. That is why the publication date and the actual time that has elapsed since I wrote the book vary so enormously.

WHAT HAS CHANGED TODAY

Since the book's first publication a number of important changes have taken place, not the least of which are my views on the confluence of science and the spiritual. It is important to note that when I wrote the book I was a callow twenty-year-old and consumed with a vision. *Mysticism and the New Physics* was the result of that vision, a product of intellectual passion, and one of the concomitants of that youthful passion was a certain amount of naïvety on my part.

In particular, as I read back over the book today I realize I was much too optimistic in my prediction that the mystical implications of the discoveries being made in the new physics indicated that some major paradigm shift was imminent in science. I still believe these discoveries will eventually precipitate a major shift in thinking, but I now realize that such a shift will take much longer than I had previously anticipated.

One of the reasons for my excessive optimism was that I did not realize how schizophrenic many physicists are when it comes to interpreting some of the new physics' more astounding findings. I thought that the mystical-sounding statements being made by such physicists indicated that they embraced a more mystical and paranormal view of the universe and this is not necessarily true. Time and experience has since taught me that some physicists are oddly schizoid when it comes to extrapolating or expanding beyond their immediate findings. They are not unlike *idiot savants*, individuals who possess a profound genius in one subject, but whose intelligence and vision is merely normal when it comes to looking beyond the narrow focus of their research.

For example, Wheeler is quite comfortable with the idea that the mind is the operative force that causes the universe to coalesce into existence. He even matter-of-factly asserts that the mind's reality-making abilities can transcend time and that an experimenter's act of observation can alter events that occurred literally billions of years in the past (for a more complete discussion of this see *Beyond the Quantum*). However, he vociferously denies the existence of psychokinesis (or, for that matter, any other kind of psychic functioning) and does not accept that one can extrapolate the participator principle into the field of the paranormal. The mind may be responsible for the creation of the entire physical cosmos, says Wheeler, but it most definitely cannot bend a paper clip without touching it.

Wheeler even condemns the use of the term *consciousness*. He prefers 'intelligent observer', which he cryptically defines as anything that is 'meaning-sensitive'. He asserts that biological entities are thus not the only things that classify as observers, and in a move that saps the participator principle of its living and organic qualities and gives it a somewhat sterile and grimly technological cast, asserts that even meaning-sensitive mechanical devices play a role in the creation of the universe. Machines can possess the magic ability to weave reality out of quantum straw, says Wheeler, but consciousness plays no role in the process.[1]

THE MOST IMPORTANT EXPERIMENT OF THE CENTURY

The paranormal implications of the new physics are not the only thing some quantum physicists eschew. Many also develop an equally bizarre schizophrenia when it comes to quantum physics' heady philosophical implications. This is strange because, since the publication of *Mysticism and the New Physics*, a number of experimental discoveries have been made that pose an even greater challenge to the Newtonian view of the universe. One of them may turn out to be the most important experiment of the century.

The experiment was performed in 1982 by physicists Alain Aspect, Jean Dalibard, and Gerard Roger of the Institute of Optics at the University of Paris, and focused on the same kind of interconnectedness that manifests between particles in the double-slit experiment. In the 1920s it was Einstein himself who first pointed out that the formulations of quantum physics predict that subatomic particles are able to communicate with one another instantaneously and regardless of the distance separating them. Einstein felt this particular prediction must be in error since an instantaneous connection between particles suggested that some faster-than-light signaling process was at work and, as we have seen, his special theory of relativity expressly forbids anything traveling faster than the speed of light. If anything other than light were able to travel that fast, said Einstein, it would imply that time travel were possible and this would open the door on all kinds of unacceptable paradoxes (one could, for instance, go back in time and kill all of one's ancestors).

The double-slit experiment is not the only situation where such instantaneous connections appear to take place. Quantum theory also predicted that some subatomic processes would produce 'twin' particles. Nearly everyone has heard stories of identical twins who allege that they are mysteriously connected and that when one is hurt the other also feels the pain. Similarly, twin particles also appear to be inexplicably connected so that anything that happens to one will instantaneously register in the other.

When Einstein first pointed out that the formulations of quantum physics implied the existence of such twin particles, the technology was not available to experimentally demonstrate their existence. In the 1970s the technology became available to offer some evidence that particles believed to be twins were actually connected, but it wasn't until 1982 that Aspect and his team settled the matter conclusively.

To do this they started by heating calcium atoms with lasers. When calcium atoms are so heated they emit pairs of photons (wave/particles of light) believed to have twin properties. For example, as anyone who has ever worn sunglasses knows, one of light's properties is its angle of polarization. Polarization is the specific angle of orientation of the light wave. According to quantum theory, the angle of a photon's polarization does not exist until it is measured. If this seems strange, one need merely remember that the country (or location) on which one's finger rests when one stops a spinning globe similarly does not exist until one actually reaches out one's finger and stops the globe.

Quantum theory also says that once the polarization of one of the calcium atom's photons has been measured, its twin will always have a related angle of polarization. Einstein felt that, if such twin photons were allowed to travel a significant distance apart, had their angles of polarization measured simultaneously, and their polarizations were indeed found to be correlated, that this would provide proof that they had instantaneously communicated (of course, Einstein felt that no such proof would ever be found).

To determine if the photons were interconnected in this seemingly faster-than-light manner Aspect and his team allowed each photon to travel through 6.5 meters of pipe and pass through special optical switches that rerouted them towards one of two polarization analyzers. The trick, and the technological stumbling block that had thwarted researchers previously, was that to rule out any unknown slower-than-light communications, the optical switches had to be able to click back and forth between the two positions every 10 billionths of a second, or faster than the roughly 30 billionths of a second it took a beam of light to cross the distance from one side of combined 13 meters of pipe to the other.

To accomplish this task, Aspect and his team used small vials of water as switches. Then they generated standing waves in the vials of water by bathing them with ultrasonic sound. When the sound was turned on and the waves were present, photons passing through the water would strike the waves and be deflected in one direction. When the sound was turned off and the waves vanished, the photons were deflected in another direction. The rapidity with which the waves could be made to appear or vanish was what enabled the vials to function as superfast optical switches.

When Aspect and his team performed the experiment and tallied the results they discovered that the angles of polarization were indeed correlated in a way that indicated the photons were instantaneously connected with one another, and this was a mind-boggling finding. It meant that some of our most cherished and accepted notions about reality are in error.

However, what was all the more astounding was that the Aspect experiment – an experiment which most assuredly changed our understanding of reality as much as the revelations of Copernicus or Darwin – went almost completely unnoticed by the mass media. Even the scientific world, as is evidenced in the response in the scientific journals, greeted it with an unusual lack of fanfare. Articles appeared announcing the results of the experiment and concluded with remarks such as 'lead(s) to realities beyond our common experience'[2] and 'indicate(s) that we must be prepared to consider radically new views of reality'.[3] But beyond that not much more has been said.

GARAGE MECHANICS AT THE EDGE OF A PRECIPICE

Why? The reasons are complex and multifaceted. One is that many physicists view talking about the implications of the Aspect experiment as belonging in the realm of philosophy and talking about the philosophical interpretation of quantum physics simply doesn't interest them. They would rather discuss data, results, and practical applications. As John Polkinghorne of the Department of Applied Mathematics and Theoretical Physics at Cambridge University puts it, 'Your average quantum mechanic is about as philosophically minded as your average garage mechanic.'[4]

Another reason is the aforementioned schizoid attitude of many physicists. For instance, the standard interpretation of quantum physics has long accepted that reality as we know it ceeases to exist at the quantum level, and many physicists accept this point of view. However, they go about their lives and research as if this finding weren't true nevertheless. As physicist Fritz Rohrlich of Syracuse University stated at a 1986 conference on quantum physics held in New York, they 'develop a somewhat schizophrenic view. On one hand they accept the standard interpretation of quantum theory, including the epistemological irreducibility of system and observer. On the other they insist on the reality of quantum systems even when these are not observed.'[5]

David Mermin of Cornell University, another physicist at the conference, agrees. In Mermin's view, physicists fall into three categories. The first is troubled by the philosophical implications of Aspect's experiment and quantum physics in general. The second is not troubled and has devised elaborate explanations to avoid worrying about it, explanations 'that tend either to miss the point entirely or to contain physical assertions that can be shown to be false'. The third group is not troubled but refuse to say why they aren't troubled. 'Their position is unassailable,' says Mermin.[6]

Yet another reason is that Aspect's findings are so staggering that even physicists who are openly bothered by its philosophical implications don't quite know what to do about them. Put another way, the French team's findings have led us to such a fog-shrouded precipice that no one is really quite sure how to start venturing into its concealing mists.

PASSION AT A DISTANCE

And what do Aspect's findings mean? Most physicists, but not all, do not believe that the interconnections between the photons are due to some faster-than-light signaling process. One reason for this assumption is that the interconnections between the particles manifest only randomly and thus do not seem capable of being utilized to send coherent communications. In other words, the two photons are somewhat like two magically interconnected roulette wheels. Every time you spin one of the magic roulette wheels and get, say, the number 7 on black, you can then predict with accuracy that when you spin the other wheel you will also get the number 7 on black. However, you have no way of controlling which number comes up on the first wheel, the numbers always come up randomly, and hence you have no way of controlling their sequence and using their interconnectedness to send a message.

As a result, many physicists assert that the propagation between the two photons is not a real signal and thus does not violate special relativity's ban on faster-than-light processes. Some, like the late Heinz Pagels of Rockefeller University, contend that we therefore do not have to worry about Aspect's findings. The instantaneous interconnections between particles is just an insignificant glitch in the fabric of reality and no further explanations are necessary. To drive his point home Pagels stressed that whatever it was that was manifesting between such subatomic particles, it should never be referred to as a 'communication' or even a 'connection', but always only as a 'correlation'. Physicist Abner Shimony of Boston University does Pagels one better and has coined the even more charming euphemism 'passion at a distance', to describe what is manifesting between such particles.

NONLOCAL FISH AND THE UNIVERSE AS A SEAMLESS WHOLE

While agreeing that the correlations between the particles are not due to some faster-than-light process, other physicists take issue with Pagel's belief that they are inexplicable 'final truths' and should therefore be ignored. One of the most notable of these is Paul Dirac, one of quantum physics' founding fathers, who believes that quantum theory is inadequate when it comes to explaining such effects and must therefore be viewed as incomplete.

Another is David Bohm who, as we have seen, has long felt that the interconnections between particles demand an entirely new concept of reality. Since the publication of *Mysticism and the New Physics* Bohm has further refined his ideas about the quantum potential and believes the correlations between the particles are due, not to faster-than-light signals, but to an even more disconcerting concept physicists call 'nonlocality'. Nonlocality is a hypothetical condition in which location ceases to exist. Bohm believes that at some level of the subatomic landscape, all semblance of location breaks down and particles such as the photons in Aspect's experiment are able to register what happens to one another, not because they are sending signals back and forth, but because their separateness is an illusion and they are actually all part of the same fundamental and cosmic unity.

To explain what he means Bohm offers the following analogy. Imagine an aquarium containing a fish. Imagine also that you are from a culture that has never seen aquariums or fish and your only knowledge about the aquarium comes from two television monitors, each connected to a separate camera, one aimed at the front of the aquarium and one aimed at its side. As you gaze at the monitors you will see two separate images of the fish, one a frontal view and one a side view. Because you have no knowledge of the true setup you might mistakenly assume that the two images represent two separate fish, two objects.

However, whenever one of the images moves you quickly notice that the other image makes a different but corresponding movement. Again, because you have no knowledge of the one true fish you might jump to the erroneous conclusion that the 'two' fish are somehow communicating instantaneously with one another, but this is not the case. At a deeper level of reality, the level of the aquarium, the two fish are actually one and the same and their apparently separate locations is an illusion. Bohm believes the same is true of the photons in Aspect's experiment. Although they look separate at our level of reality, at a deeper nonlocal level of reality, a level analogous to the aquarium, they, and indeed all particles in the universe, are all aspects of a deeper fundamental unity.

What implications does this fundamental unity have for our level of existence? For one, Bohm believes it means that when we try to divide the universe up into things like 'electrons', 'photons', and so on, we are only performing an abstraction. 'Out there' the universe is always a seamless and indivisible whole and hence electrons exist only as ideas in our minds. As I noted in the original introduction to *Mysticism*, more and more the new physics seems to underscore the fact that the world *does not* yield to us directly. A description of the world always stands in between, and all too often when we think we are analyzing a phenomenon, we are really only analyzing a concept and therefore the use of a word.

USING QUANTUM INTERCONNECTEDNESS AS AN INTERGALACTIC TELEGRAPH SYSTEM

Is it possible to somehow bypass the random nature of the correlations between subatomic particles and use their apparent nonlocal interconnectedness to send coherent messages? Although this is a highly controversial point and most physicists believe it cannot be done, a few still see it as a possibility. Bohm, for one, thinks that Aspect's findings strongly suggest that there are still deeper levels of reality beyond the subatomic landscape and feels that these deeper strata of existence may contain processes that transcend the speed of light limit. 'So long as the present type of experiment is done, the theory of relativity will still be saved,' says Bohm. 'But if we could manage to get deeper than that then we might find that there was something faster than light.'[7]

Physicist Brian Josephson, who won a Nobel prize in 1973 for his work on quantum tunneling and superconductivity, also feels that Aspect's findings open the door on faster-than-light communications and 'raises the possibility that one part of the universe may have knowledge of another part – some kind of contact at a distance under certain conditions'.[8]

Even more controversially, in 1987 physicists Dipankar Home of the Saha Institute of Nuclear Physics in Calcutta, Amitava Raychaudhuri of the University of Calcutta, and Amitava Datta of the University of Dortmund, West Germany, announced that they had found a way to circumvent the statistical static of the quantum interconnectedness of all things, a way that might someday allow us to use such interconnectedness to send messages.

The method they proposed involves a rare subatomic particle called a kaon. Home's group asserts that they have found reason to believe that – unlike other twin particles, which respond uniformly to measurements performed on them – the statistical properties of one kaon in a set of twins depends on what is done to the other particle. 'If you don't make any measurement, you get one result. But if you do make a particular measurement, you get another result,' says Dr Datta.[9] Thus, by alternately measuring or not measuring one kaon in a set of twins, one can send a signal that will register in the statistical properties of the other particle.

Shimony believes the team has made an error in their mathematical calculations. Even physicists who cannot instantly spot an error in the mathematics believe that such a flaw must exist and that Home and his team cannot possibly be correct. For the moment the jury remains out and only further study will reveal whether Home's group has found a way around the statistical static of quantum interconnectedness.

OTHER IMPORTANT DEVELOPMENTS: A CHAPTER BY CHAPTER UPDATE

CHAPTER ONE Observer and Participant

The Death of the Clockwork Universe

Aspect's experiment is not the only important development that has taken place since the first publication of this book. There have been many others. For instance, in chapter one I noted that one of the most troubling issues facing the founding fathers of quantum physics was the realization that many quantum events are indeterministic and cannot be predicted with certainty. This was disturbing because scientists had long assumed that, given enough knowledge about a system, its future states could be determined with accuracy. One of the most frequently cited indications that this was so, and that the workings of Mother Nature were as predictable as the movements of a clock, was the striking success with which Newtonian physics was able to predict the motions of the planets.

Recently, however, it was discovered that even the motion of the planets is not quite as clockwork as was previously assumed. In 1989 French mathematician Jacques Laskar fed information about the current positions and orbits of the planets of the Solar System into a high-speed computer. His intention was to see what the movement of the planets would look like over the next 200 million years. To his surprise, he discovered that the orbits of the inner planets, including the Earth, lose all predictability in only 10 million years – a mere blink of the eye in cosmological terms. This doesn't mean that the planets are going to start bumping into one another, says Laskar. They do stay within certain bounds. But after 10 million years small variations in their orbits cause them to become quite chaotic within those bounds. By making such a discovery Laskar appears to have placed the final nail in the coffin of the clockwork universe.

Trying to Steal a Peek at Schrödinger's Cat

As we have seen, one of the most disturbing aspects of the new physics is the idea that an observer actually changes the reality of a subatomic particle by observing it. The extraordinariness of this is underscored by Schrödinger's cat, a thought experiment in which a cat in a sealed box must be viewed as both alive and dead until an observer opens the box and looks at the cat – an observational act that is, according to quantum theory, the magical wave of the wand that causes the actual fate of the cat to finally coalesce into existence. For seven decades now physicists have sought ways to dispel this disconcerting state of affairs, and to, in a manner of speaking, steal a peek at Schrödinger's cat while it is still sealed in the box. So far all have failed and have only further cemented the idea that the act of observation does indeed play a role in the creation of reality.

One of the most interesting of these recent attempts involved an experiment performed in 1989 by physicist Wayne Itano and his colleagues at the National Institute of Standards and Technology in Boulder, Colorado. The group bathed beryllium atoms with radio waves, a process that would gradually cause the atoms to enter a more excited and energized state. As they did this they also periodically 'looked' at the beryllium atoms by shooting a short pulse of laser light at them. Excited beryllium atoms are invisible in laser light, but unexcited beryllium atoms glow brightly. Thus, bathing the beryllium atoms with laser light was somewhat like turning on a light in a microwave oven to see how many popcorn kernels have popped, and allowed Itano's team to determine how many beryllium atoms had entered an excited state and how many had not.

Itano's team discovered that if they waited a quarter second before turning the laser on, nearly all of the atoms had reached the excited state. If they bathed the atoms in laser light halfway through the quarter-second interval (or one-eighth second into the experiment), by the end of the quarter second only half the atoms had become excited. If they bathed the atoms with laser light four times during the quarter-second period only one-third of the atoms were excited by the end of the allotted time. If they shot 64 pulses of laser light at the atoms, virtually none of the atoms made the transition.

This was a most peculiar state of affairs. Imagine a pot of water sitting on a burner on the stove that literally *would not boil as long as you kept turning the kitchen light on and looking at it!* This is in essence what Itano's team discovered. Itano's explanation is that, like Schrödinger's cat which is always both alive and dead, the beryllium atoms are always in both states simultaneously, always both excited and unexcited. However, they cannot manifest this way when we look at them, and an act of observation via a pulse of laser light forces them to assume only one reality or the other.

In addition, it takes a little time for them to make the transition from an unexcited state to an excited one. Hence, they are not unlike the frog in the fairy tale that climbs up the side of a well three feet every night, but slips back down two feet every morning when it is struck by the first light of dawn. If you shine a pulse of laser light at a beryllium atom and look at it before it has completed its transition it has no choice but to collapse one of its realities and return to an unexcited state, says Itano. Like shy children who will not get dressed while they are being watched, beryllium atoms will not complete their transition while they are watched, and if you look at them frequently enough you keep them forever frozen in their unexcited state. 'It's an effect that people have talked about for years,' says Itano, 'but it was hard to show in a clean and simple way that it actually happens.'[10]

The effect uncovered by Itano's team cannot be explained in terms of classical physics, but offers additional evidence of the eerily powerful effect the act of observation has on subatomic phenomena.

CHAPTER TWO *A Holographic Model of Consciousness*

Anticipating Sheldrake's Morphogenetic Fields

In chapter two I suggested that the inexplicable interconnections between various portions of the human brain and the mysterious interconnections between subatomic particles are related and that both may be due to Bohm and Hiley's proposed 'quantum potential', an as yet undiscovered field that has the ability to cut across both space and time. I further proposed that the quantum potential might also play a role in the organization of biological systems in general and help explain some of the unsolved problems of embryological development.

In 1981 Cambridge biologist Rupert Sheldrake published his now famous *A New Science of Life*, a book in which he proposed exactly the same thesis. In the book Sheldrake called the time and space-transcending fields that organize life and govern embryological development 'morphogenetic fields'. In a more recent book titled *The Presence of the Past* Sheldrake even draws an explicit connection between morphogenetic fields and Bohm's ideas on the interconnectedness between subatomic particles.[11] (It should be noted that I am not trying to take credit for Dr Sheldrake's ideas. Since our thoughts on the matter were both initially published the same year it is clear that each of us arrived at them independently. But I am gratified that an hypothesis I formulated as a teenager reading and thinking in relative isolation in the early 1970s has found support in a thinker as eminent as Dr Sheldrake.)

What is important is that Sheldrake has also proposed ways that the existence of such fields might be experimentally proven. For example, Sheldrake believes that morphogenetic fields also govern behavior and that the more a species performs a certain task, the easier it becomes for future members of the species to learn and perform the same task. In one experiment conducted in 1984 and designed to test his hypothesis a picture containing a concealed image of a face was shown by the BBC to an estimated 80 million viewers. As Sheldrake predicted, of 6,500 individuals tested, those who saw the picture after it had been deciphered by 80 million people found it slightly easier to see the hidden face than those who were confronted with the picture before the airing. Such results are of course inconclusive, but they are compelling and will hopefully inspire further study.

An Explosion of Holographic Ideas

Since the publication of *Mysticism and the New Physics* the holographic idea has also enjoyed a certain amount of popularity. Although it is still a highly controversial theory, over the past decade and a half researchers from an astonishingly wide range of different fields have used it to explain a host of puzzles in their respective fields. For example, in 1985 Dr Stanislav Grof, chief of psychiatric research at the Maryland Psychiatric Research Center and an assistant professor of psychiatry at the Johns Hopkins University School of Medicine published a book in which he concluded that existing neurophysiological models of the brain are inadequate and that only a holographic model – and the interconnectedness between all things it implies – can explain the mind's ability to access archetypal, racial, and even genetic information from the collective past of the human race.[12]

In a 1987 book titled *Synchronicity: The Bridge Between Matter and Mind*, physicist F. David Peat of Queen's University in Canada says that the same holographic interconnectedness of all things may also explain synchronicities.[13] In the 1980s University of Connecticut psychologist Kenneth Ring proposed that even near-death experiences may be explained by the holographic model of consciousness and that the dazzling light-filled realm individuals visit during such experiences are real and are actually just different channels on the cosmic television set we call reality.[14] Indeed, so many are the areas in which the holographic idea is being applied that I recently wrote an entire book on the subject (the aforementioned *The Holographic Universe*).

CHAPTER THREE *Superspace*

New Ripples in the Quantum Foam

Wheeler's belief that elementary particles consist of microscopic wormholes threaded by electric or other field lines has not proved to be a useful description, but several of his related ideas have. One is the notion that the universe is permeated with a Swiss cheese of wormholes, an idea that again challenges our commonsense understanding of location. Although it appears doubtful that such wormholes explain the nonlocal effects that take place between subatomic particles, some physicists believe that Wheeler's ideas have at least provided fertile soil in which to sprout ideas. One such physicist is Peat who wrote recently that Wheeler's wormhole concept helped 'shatter the hypnotic hold that local reality holds over us and may encourage physicists to think in new ways'.[15]

The waning of Wheeler's notion that particles are lines of force trapped in space-time notwithstanding, the notion that space-time is actually a seething quantum foam of wormholes has also survived. In fact, in the past decade there has been an explosion of interest in the concept and many believe accepting the existence of wormholes sheds light on some other unsolved puzzles.

One is the mystery of how the entire universe came into being. As we have seen, many scientists believe that blackholes are apertures to some absolute elsewhere, a region so beyond space-time that we can say nothing about its properties. Because blackholes are believed to be a region where all of space-time collapses to a single point, they are sometimes called 'singularities'.

For more than two decades now scientists have believed that the universe itself erupted out of a singularity in a cosmic explosion known as the Big Bang. The Big Bang theory also holds that our current universe should still be salted with a great number of singularities. The only problem is that it is difficult to describe how the universe erupted out of a region which we have no concepts to describe. As the famous Cambridge physicist Stephen Hawking puts it, 'We cannot predict what comes out of a singularity. It is a disaster for science.'[16]

Hawking has worked on the physics of singularities for almost three decades and has been a longtime supporter of the idea that our current universe was born out of a singularity. It thus came as quite a shock when he announced at a 1987 Workshop on Quantum Cosmology held at the Fermi National Accelerator Laboratory in Batavia, Illinois, that he had abandoned his long held position. 'I have changed my mind,' he said simply.

Hawking now believes that it is wormholes, not singularities that have played a major role in the creation of the universe. He further believes that rather than being a doorway to some absolute elsewhere, the centers of blackholes are actually wormholes into little universes, 'baby universes' that are somewhat like pouches in the fabric of space-time.

Hawking believes the existence of wormholes helps solve another longstanding puzzle, the problem of how to explain Einstein's *cosmological constant*. In 1917 Einstein attempted to use his newly developed general theory of relativity to describe the shape and evolution of the universe. At the time the universe was believed to be static and unchanging. However, armed with the powerful new understandings of relativity Einstein realized that if this were true, the force of gravity should have long ago caused the universe to collapse. To explain why the universe had not collapsed Einstein asserted that there must be some kind of antigravitational force at work in the cosmos, a force that prevented the universe from being crushed under its own gravitational weight. He called this antigravitational force the cosmological constant.

In the 1920s, however, astronomer Edwin Hubble made his now famous discovery that the universe is expanding. With this revelation everyone, including Einstein, realized that it was the expansive force – believed to be a result of the aftermath of the Big Bang – that was preventing the gravitational annihilation of the universe, and the cosmological constant was no longer required. Einstein quickly amended his ideas and admitted that his theorized cosmological constant was the biggest blunder he ever made. Yet, decades later, physicists have had a remarkably difficult time getting rid of this cosmological bugaboo.

The problem is that, like a bubble under the wallpaper, every time physicists try to get rid of the cosmological constant it crops up somewhere else. For instance, once physicists accepted the Big Bang theory and the idea that universe began as a single and homogeneous primordial mass – regardless of whether it erupted from a singularity or a wormhole – that left them with the problem of explaining how this primordial homogeneity evolved into the multiplicity of different particles and forces we see today. To solve this problem physicists have spent the last several decades trying to come up with a single mathematical supertheory that would unify all the known forces of nature.

It has been found that one simple and mathematically elegant way to do this is to propose the existence of several entirely new fields. For example, in the late 1970s physicists Sheldon Glashow, Abdus Salam, and Steven Weinberg were able to show how the electromagnetic and weak nuclear force (the force responsible for radioactivity) could be unified by proposing the existence of a field called the 'Higgs boson field'.

Unfortunately, accepting the existence of these new fields brings a price with it. If the universe is permeated by such fields it means that seemingly empty space is not empty at all but contains an enormous amount of hidden energy. And as it turns out, this pervasive sea of latent energy behaves strikingly like Einstein's cosmological constant.

How, then, are we to rid ourselves of this cosmic albatross? Hawking's mathematical formulations made him realize that if the cosmological constant existed and were large enough for astronomers to measure, it meant that the probability of the universe evolving to its current state was vanishingly small. But the mathematics also revealed that if space-time were actually an effervescent landscape of quantum foam it made our universe a much more likely outcome of the Big Bang and eliminated the cosmological constant once and for all.

In 1988 physicist Sidney Coleman of Harvard tossed his hat into the ring of quantum foam believers and proposed a similar view. One of the problems with all unification theories, said Coleman, is that they continue to view space-time in strictly Einsteinian terms, almost like a three-dimensional sheet of rubber. But if space-time were more like a sea of quantum foam, the wormholes permeating such a foam would make ironing out the problems in such unification theories much easier.

Coleman noted that his calculations revealed that large wormholes, wormholes the size of baseballs, basketballs, and planets, would be exceedingly rare. But wormholes on the subatomic scale would be everywhere. In fact, there should be trillions and trillions of them winking in and out of existence every second in the space occupied by our thumbs. According to Coleman these wormholes are not unlike voracious amoebae and would gobble up all the latent energy of the fields that permeate space, such as the Higgs boson field, and comprise the cosmological constant. Thus, in both Hawking's and Coleman's formulations, the quantum foam – a phenomenon of the world of the very small – appears to help us understand the world of the very large, the origin and evolution of the cosmos itself.[17]

CHAPTER FOUR *Beyond the Light Cone* and CHAPTER FIVE *The Shape of Time*

The Tachyon may be Dead, but Time Travel is Not

Over the last ten years enthusiasm for the existence of faster-than-light particles or tachyons has waned considerably. The reason is simple. In 1934 the Russian physicist Pavel Cerenkov discovered that if an electrically charged particle were to travel faster than the speed of light, it would create the light equivalent of a sonic boom and itself give off light – a luminescence now referred to as Cerenkov radiation. Thousands of photographs of high-energy collisions are taken every day in dozens of laboratories around the world, and not one has ever revealed any evidence of a Cerenkov radiation trail that would betray the existence of a tachyon.

Of course tachyons may not have an electrical charge, but even this possibility appears to have been ruled out. The production of a particle moving faster than light would require no small amount of energy and our instruments should be able to detect this too. But again, exhaustive searches of data from both high-energy collisions in accelerator laboratories and from cosmic ray detectors aimed at the heavens have failed to turn up any compelling evidence of the type of energy deficits that tachyons should produce. As Columbia physicist Gerald Feinberg, who also once proposed a tachyon theory, has joked, the only place one can find a tachyon today is in the dictionary.[18]

Interestingly, although the tachyon idea has faded, theories about breaking the time barrier have flourished. For instance, recently physicist Yakir Aharonov of the University of South Carolina pointed out that quantum theory itself suggests that it may be possible to create a 'quantum time-translation machine', a device capable of taking an event in the present and moving it into the future. Aharonov's device would not be practical to build because it requires that one enclose a quantum system in an extremely massive sphere whose radius can be changed at will, and also relies on notorious capricious and unpredictable quantum processes to operate, but it does at least provide a theoretical way to construct a kind of time machine.[19]

In 1985 Michael Morris and Ulvi Yurtsever, and their Ph.D. thesis adviser Kip Thorne of the California Institute of Technology, started wondering if large-scale wormholes could be used for space travel and even time travel. By delving into the mathematics of wormholes the Caltech team discovered that the laws of physics do indeed appear to permit wormholes to be used for this purpose. They found that to turn a wormhole into a time machine one need merely move its mouth, a feat that could be accomplished by giving the wormhole an electric charge and using electric fields to move it. Once this is done, says the team, a spaceship could enter the stationary mouth and exit through the moving mouth at a point in time *before* one entered. The team believes that by positioning the mouths close to one another it may even be possible to knock a ball through such a hole, have it travel back in time, and come out and knock itself back through the hole again.

Thorne and his colleagues believe it may also be possible to use one of the very tiny wormholes in the quantum foam for this purpose. To keep such a wormhole from snapping shut a scant instant after it appears, the team suggests manipulating its electromagnetic field by putting plates of gold or copper at opposite ends of its mouths. The feasibility of using wormholes for time travel rests on one issue, say the physicists, whether or not a theoretical relationship called the 'averaged weak energy condition' can be violated. If it can be violated, and Thorne and his colleagues believe it can, then wormholes could hypothetically be used as doorways into other regions of time.

As to whether this would allow someone to, say, travel back in time and kill one's ancestors, the team admits ignorance. 'We're not forced to confront any philosophical implications because this is theoretical physics, not philosophy,' says Thorne. 'It may turn out that the averaged weak energy condition can never be violated, in which case there could be no such things as traversible wormholes, time travel or a failure of causality.' And then with subtle irony he adds, 'It's premature to try to cross a bridge before you come to it.'[20]

CHAPTER EIGHT *The Reality-Structurer*

The World as a Construct of Consciousness

Several developments have taken place that have a direct bearing
on the idea that the mind can interact with the material world. In a
1986 article published in the *Journal of the American Society for Psychical Research* Bohm announced that although he is not certain
psychokinesis exists, he feels that quantum interconnectedness at
least provides us with a way of understanding how it could.

While cautioning that all concepts exist in our heads and do not
exist 'out there', Bohm none the less posits that we can look at
reality as if it consists of two levels. He calls the level we inhabit –
where things like electrons, toaster ovens, and human beings appear
to be separate from one another – the *explicate* order. The level of
subatomic reality – where things cease to have separate location,
quantum interconnectedness reigns, and all things become a seamless
and unbroken whole – he calls the *implicate* order.

As we have seen, because everything in the universe is ultimately
constituted out of things that exist at this unbroken level, the apparent
separateness of objects at our own level of existence is also an
illusion. Hence, Bohm feels it is as meaningless to talk about the
consciousness of an observer interacting with the subatomic particle
being observed as it is to talk about two figures in a Rembrandt
painting interacting. Because we are ultimately not separate from the
objects we observe, and at some level are a continuum with them,
we cannot interact with them. In some strange sense we *are* them.

However, this conclusion on Bohm's part does not mean he is
against bringing the concept of consciousness into our understanding
of physics. In fact, it leads him to some rather extraordinary conclusions about consciousness. For example, because we are constituted
out of this nonlocal level Bohm feels it is ultimately meaningless to
speak about consciousness as having a specific location. It may
manifest inside our heads while we function in life, but the true
home of consciousness is in the implicate, says Bohm. Thus, consciousness, the great ocean of consciousness that has divided itself up
into all human beings, also exists in all things. Despite its apparent
inanimate nature, in its own way a rock is also permeated with
consciousness. So are grains of sand, ocean waves, and stars.

Bohm also believes it is clear that consciousness can interact with physical matter in that we can 'think' we want to move a finger and our finger will move. Thought is moving the finger, says Bohm. As a result it is not such a great leap to think that the right 'resonance' of thought might also reach out from our minds via the infinite interconnectedness of the implicate and cause a seemingly inanimate object like a rock to move. Crudely speaking, all things are extensions of our bodies, and like some vast and meandering underground stream, consciousness pervades all things.[21]

Another physicist who has taken the idea of psychokinesis even further is Robert G. Jahn, a professor of aerospace sciences and dean emeritus of the School of Engineering and Applied Science at Princeton. For over a decade now Jahn has been carefully accruing actual experimental evidence that the mind can interact with physical matter. For instance, in one lengthy series of experiments Jahn had people sit in front of a random number generator – a kind of automatic coin flipper – and try to will the machine into producing more heads than tails. Over the course of literally hundreds of thousands of trials he discovered that volunteers could indeed exert a small but statistically significant effect on the random number generator's output.

In another lengthy series of experiments Jahn had people sit in front of a vertical pinball-like device and try to influence the direction that three-quarter-inch marbles fell. Again after many thousands of trials Jahn obtained small but statistically significant evidence that his volunteers were able to mentally influence which way the balls fell. In a 1987 book written with his colleague Brenda J. Dunne and titled *The Margins of Reality: The Role of Consciousness in the Physical World* Jahn detailed his research and states, 'While small segments of these results might reasonably be discounted as falling too close to chance behavior to justify revision of prevailing scientific tenets, taken in concert the entire ensemble establishes an incontrovertible aberration of substantial proportions.'[22]

What is the explanation? Jahn and Dunne believe that it is the interaction between consciousness and reality described in quantum theory that provides the key, and it is time we accept this. They propose that since all quanta can manifest either as a wave or a particle, it is not unreasonable to assume that consciousness does as well. It is particle-like when it appears to be inside our heads, but in its wave-like phase it can interact with the physical world. However, like Bohm, they do not believe that it is meaningful to speak of consciousness as existing separately from the physical world. 'The message may be more subtle than that,' says Jahn. 'It may be that such concepts are simply unviable, that we cannot talk profitably about an abstract environment or an abstract consciousness. The only thing we can experience is the interpenetration of the two in some way.'[23]

Controversial as both Bohm and Jahn and Dunne's points of view are, they are significant because they advance science into at least a modest acceptance that we are reality-structurers not only in a philosophical sense, but literally, and in a way that has stupendous ramifications for our understanding of ourselves and our role in the universe.

CHAPTER NINE *The New Cosmology*

Will the Mind-based Nature of Reality ever be Accepted?

As has been mentioned, one of the most troubling aspects of the new physics is that so little has been said or done about its most astounding assertion – that the mind plays a role in the creation of the material universe. In commenting on this fact at the 1987 Loyola Conference on Mathematical and Interpretational Problems in Relativistic Quantum Theory, physicists T. Gornitz and C. F. von Weizsäcker noted that even quantum physics' founding fathers 'fell into stammering' when they were asked to discuss the implications of this point of view.

Judging from the talks delivered at the conference, it is a situation that is not going to be cleared up in the near future. Although some physicists advocated more openly recognizing the role consciousness plays, others continued to dismiss its importance. For example, Gornitz and Weizsäcker want to bring the observer more explicitly into our understanding of quantum physics and have tried to formulate meaningful ways to relate quantum phenomena to states of mind. 'There is no distinction between substances called mind and matter,' says Weizsäcker.[24]

Others at the conference voiced strenuous objections to this idea. Edward Teller of Lawrence Livermore Laboratory typifies this point of view when he says, 'I have some concept of looking at myself as an object. There is a difference between looking at myself as a piece of matter and looking at myself as a spirit or mind.' John G. Cramer of the University of Washington in Seattle, another physicist at the conference, announced that he was trying to formulate a new interpretation of quantum physics that denies the role of the observer altogether.[25]

Given findings such as Itano's, that beryllium atoms will not pass into a more energized state while they are being looked at, it seems difficult to deny that the mind plays some role in the creation of reality. It is discouraging that nearly a century after it was first proposed, the idea that consciousness helps weave the warp and weft of the world is still being so hotly contested. Such resistance leads one to ask if the mind-created nature of reality will ever be accepted?

Despite enormous resistance to the idea in many quarters, there continue to be signs that a paradigm shift is slowly taking place, that a new cosmology is gradually dawning. In addition to individuals like Bohm, Jahn, and the other scientists mentioned above a number of other eminent researchers have also added their names to the ranks of believers.

One is Roger Sperry of the California Institute of Technology. In 1981 Sperry won a Nobel Prize for his pioneering split-brain studies on the brain's left and right hemispheres. As a result of the award he was invited to write the lead article for the *1981 Annual Review of Neuroscience*. Recipients of this honor usually write a review of the past year's accomplishments in a specific area of research. Sperry did something very different. In an article titled 'Changing Priorities' Sperry announced that, after spending a lifetime studying the brain, he had become disillusioned with the materialist and behaviorist doctrine that has dominated neuroscience for the better part of this century. After long and careful thought he had come to the conclusion that science should not only stop disregarding consciousness, but recognize its extraordinary importance in the scheme of things. 'Instead of renouncing or ignoring consciousness . . . (we should) give full recognition to the primacy of inner conscious awareness as a causal reality.'[26]

Sperry is not the only Nobelist to make such a statement. At a 1986 colloquium on the 'Unsolved Problems in the Science of Life'

Nobel prize-winning biologist George Wald announced that as he nears the end of his life as a scientist he has been forced to make a similar reassessment:

A few years ago it occurred to me that ... I had always thought of consciousness, or mind, as something that required a particular complex central nervous system and was present only in the highest organisms. The thought now was that mind had been there all the time, and the reason this is a life-breeding universe is that the pervasive, constant presence of mind had guided the universe that way ... Our growing scientific knowledge ... points unmistakably to the idea of a pervasive mind intertwined with and inseparable from the material universe. This thought may sound pretty crazy, but such thinking is millennia old in the Eastern philosophies . . .[27]

Psychiatrist Grof, who is one of the founders of Transpersonal Psychology, also supports such an idea. He believes that if one makes an honest assessment of quantum physics, consciousness research, neurophysiology, and even ancient and Oriental spiritual philosophies, shamanism, and parapsychological phenomena, one cannot help but come to the conclusion that consciousness can 'modify phenomena' in the material world. 'It seems obvious that we are approaching the time of a major paradigm shift,' says Grof.[28]

These are only a few of the voices that have recently been raised. There are others. In his book *The Reenchantment of the World* Morris Berman, a history of science professor at the University of Victoria, asserts that science will remain inadequate as long as it attempts to describe Nature only from the outside. He agrees with philosopher Peter Koestenbaum when he says that 'there is no specific border in which mind becomes matter'. Building on this thought, Berman argues compellingly that Nature is only revealed through our relations with it, and it is crucial that we become cognizant of this 'participatory' factor if we are to truly fathom the wonders of the cosmos.[29]

In his 1989 book, *Recovering the Soul* Larry Dossey, M.D., a former chief of staff of Medical City Dallas Hospital, also argues passionately in favor of 'the absolute status of human consciousness'. That is, 'consciousness as fundamental and not derivative of the physical; consciousness as infinite in space and time'. Dossey is particularly taken with Bohm's ideas about nonlocality and believes that investigating this mysterious nether region will ultimately help us establish that consciousness is indeed timeless, spaceless, and immortal. '*Recovering the nonlocal nature of the mind . . . is essentially a recovery of the soul,*' says Dossey.[30]

And in his groundbreaking *Global Mind Change* Willis Harman, a psychologist, engineer, and former senior social scientist at Stanford Research Institute International, spends an entire book examining developments in physics, parapsychology, psychology, and sociology, and pointing out why they indicate that both science and civilization in general are in the midst of a major shift from what he calls an 'M-1 metaphysic', or the belief that the basic stuff of the universe is matter-energy, to an 'M-3 metaphysic', or the belief that the ultimate stuff of the universe is consciousness. 'Mind or consciousness is primary, and matter-energy arises in some sense out of mind,' says Harman.[31]

Such remarks, especially ones made by such eminent thinkers, do indeed suggest that ripples of change are passing through our collective intelligence. However, having already been wrong once in assuming that such a state of affairs indicates that a major paradigm shift is imminent, I am not going to make the same error again. As I stated earlier, I agree with conclusions such as Harman's that the findings of the new physics *are* shifting us inexorably into a new metaphysic. Only now I realize that it is a shift that will take some time. When one wants to train the trunk of a miniature *bonsai* tree to turn in a new direction, one cannot simply bend it in that direction all at once or the trunk will break. One must wrap the trunk with wire and slowly, ever so slowly, shape it in small increments. Only then can it survive the stress.

Living things resist precipitous transformation. Organic things – from cellular organisms to the belief systems that take root in the fertile soil of our thoughts – prefer homeostasis. In a living universe, a universe constantly awash with restless and dynamic forces, resistance to change even has a certain amount of survival value. That is why change often takes such a long time. But in a dynamic universe, change is also inevitable, and this too is how it must be. It is like growing *bonsai*.

Glossary of Scientific Terms

algorithm – Special step-by-step method of solving a kind of mathematical problem.

alpha particle – Positively charged particle consisting of two protons and two neutrons.

biograviton – Conjectured gravitational field composed of massive gravitons, on a scale of approximately 10^{-4} cm, which control the holistic organization of life.

blackhole – Pucker in the fabric of space-time with a gravitational field so intense that nothing, even light, can escape from it.

Brownian movement – Constant zigzag movement of particles in a liquid or gas.

causality – Interrelation or connection between cause and effect.

cognitional multi-dimensional projection spaces – John A. Lilly's postulated portions of the human mind which can synthesize entire inner realities.

complementarity – Theory, proposed by Niels Bohr, that particles in microscopic systems behave simultaneously as waves and as particles.

configuration space – Illustrative diagram in which an object is depicted in time as well as in the three dimensions of space.

contingent – Dependent upon statistical limits, not absolute.

curvature – Geometric property of three-dimensional space.

cybernetics – Science of control and communication within and between machines, animals and organizations. Specifically, the study of the common principles in the workings of computing machines and those of the human nervous system.

depolarize, to – To destroy or counteract the polarization of.

determinism – Belief that all effects are created or determined by absolute causes.

diffraction – Scattering or breaking up of waves or particles.

elsewhere – The 'something' or region which exists outside the light cone (q.v.), literally beyond space-time.

entelechy – Hypothesized biological organizing principle above and beyond the genetic design of life.

ether – Postulated invisible substance that pervades space and serves as the medium for the transmission of light waves and other forms of radiant energy.

Euclidean geometry – Basic geometry devised by Euclid, *c.* 300 B.C.

field – Space in which electric, magnetic, or dynamic lines of force are active.

geometrodynamics – Branch of quantum physics developed by John A. Wheeler in which matter is seen as being composed entirely of curvature; study of the geometrical structure of space.

gravitons – Proposed wave/particles of gravity.

half-life – In nuclear physics, the time required for disintegration of half the atoms in a quantity of a radioactive substance.

holism – view that an organic or integrated whole has an independent reality which cannot be understood simply through an understanding of its parts.

hologram – Three-dimensional photographic image developed with a laser.

indeterminism – Belief that some effects literally have no causes.

ion – Electrically charged atom or radical.

L-fields – Dr Harold Saxton Burr's proposed dynamic fields which holistically organize life.

light cone – Illustrative diagram which depicts each point in space as existing in time, but also takes into account the elsewhere regions which are literally beyond space-time.

metaprinciple – Principle which goes outside a closed theory for the 'decidability' of propositions.

metatheorem – Theorem or postulate which goes outside what is known about a system for the 'decidability' of propositions.

miniblackholes – Bubbles in the quantum foam 10^{-33} cm in diameter. They form the mouth of a wormhole in which a miniwhitehole forms the other mouth.

miniwhiteholes – See miniblackholes.

neutrino – Neutral particle smaller than a neutron.

neutron – Subatomic particle having no electric charge and a mass very close to that of a proton.

omnijective – Pertaining to the belief that consciousness and the physical world are not separate, but form one fundamental arena of awareness which is omnijective as opposed to being subjective or objective.

operator – In quantum theory, any determined mathematical function which affects the wave function of a particle.

participator – Experiencer who not only observes an occurrence but changes it by the mere act of observation.

photon – Discrete unit of electromagnetic energy.

polarization – Producing or acquiring of the condition of having magnetic poles (negative and positive).

potential difference – Difference in electric potential or charge.

proton – Subatomic particle with positive electric charge; one of the two primary constituents of the nucleus.

psychoenergetic – Acting under or possessing energy created by the human consciousness.

psychokinetic – Moved by the power of the human consciousness.

quantum – Discrete unit of energy.

quantum foam – John A. Wheeler's proposed picture of space as composed of microscopic bubbles forming what can be conceptualized as a carpet of foam.

quantum interconnectedness – Proposition that through the bubbles in the quantum foam all points in space and time are connected to all other points in space and time.

quantum mechanical – Having to do with the workings or mechanics of atomic systems.

quantum physics – Branch of physics which deals with the study of atomic systems.

quantum potential – Connecting principle between quantum mechanical events which exists literally beyond space-time.

quantum principle – View that the explanations of quantum theory lie outside quantum theory.

quantum transition – Any quantum mechanical occurrence.

reality-structurer – Portion of the human consciousness which affects matter-space-time.

retrocausal – Phenomenon in which the effect temporally precedes the cause.

self-reference cosmology – View that the universe exists ultimately as an unbroken whole in which all parts are simultaneously creating and being created by all other parts.

superspace – John A. Wheeler's view that space is composed of quantum foam and that all the matter in the universe is composed of this one ultimate substance.

tachyon – Hypothetical faster than light particle.

vector – Quantity, such as a force or velocity, having direction and magnitude; line representing such a quantity.

virtual – Existing only in effect or essence, but not in actual fact or name.

wave function – Abstract line or function in configuration space representing the physical state of a system.

world line – Abstract line in configuration space representing a system's position both in three-dimensional space and in time.

wormholes – Holes in the quantum foam which interconnect all regions of space-time.

Notes

1. Zsolt Aradi, *The Book of Miracles*, Farrar, Straus and Cudahy: New York, 1956.
2. Solomon E.. Asch, 'Opinions and Social Pressure', *Scientific American*, vol 193, no 5.
3. W. Ross Ashby, *Design for a Brain*, Wiley: New York, 1952.
4. Michael Audi, *Quantum Mechanics*, University of Chicago Press: Chicago, 1973.
5. Sri Aurobindo, *The Hour of God*, Sri Aurobindo Ashram Press: Pondicherry, 1959.
6. ———, '*The Life Divine*', *Arya* (August 1914–January 1919), Sri Aurobindo Ashram Press: Pondicherry.
7. ———, *Savitri: A Legend and a Symbol*, Sri Aurobindo Ashram Press: Pondicherry, 1954.
8. ———, *The Synthesis of Yoga*, Sri Aurobindo Ashram Press: Pondicherry, 1948.
9. Ludwig von Bertalanffy, *General Systems Theory*, George Braziller: New York, 1968.
10. Olexa-Myron Bilaniuk and E. C. George Sudarshan, 'More about Tachyons', *Physics Today*, December 1969.
11. ———, 'Particles Beyond the Light Barrier', *Physics Today* 22, no. 5 (1969): 43.
12. John Blofeld, *The Tantric Mysticism of Tibet*, E. P. Dutton: New York, 1970.
13. ———, *The Way of Power*, George Allen & Unwin: London, 1970.
14. ———, *The Zen Teaching of Hui Hai*, Rider: London, 1969.
15. R. H. Blyth, *Games Zen Masters Play*, edited by Robert Sohl, and Audrey Carr, New American Library: New York, 1976.
16. D. Bohm and B. Hiley, 'On the Intuitive Understanding of Non-Locality as Implied by Quantum Theory', preprint February 1974, available from authors, University of London, Birkbeck

College, Malet St., London. (As quoted in Jack Sarfatti, 'Implications of Meta-Physics for Psychoenergetic Systems', *Psychoenergetic Systems*, vol. 1, Gordon and Breach: London, 1974.)

17. Jorge Luis Borges, *Ficciones*, Grove Press: New York, 1962.

18. John Brockman, *Afterwords*, Anchor Books: New York, 1973.

19. Jerome S. Bruner, *On Knowing, Essays for the Left Hand*, Belknap Press: Cambridge, 1962.

20. Harold Saxton Burr, *The Fields of Life*, Ballantine Books: New York, 1972.

21. Fritjof Capra, *The Tao of Physics*, Shambhala Publications: Berkeley, 1975.

22. Carlos Castaneda, *Journey to Ixtlan*, Simon and Schuster: New York, 1972, and *Tales of Power*, Simon and Schuster: New York, 1974.

23. Garma C. C. Chuang, *Teachings of Tibetan Yoga*, Citadel Press: Secaucus, N.J., 1974.

24. Arthur C. Clarke, *Childhood's End*, Ballantine Books: New York, 1953.

25. Olivier Costa de Beauregard, 'Time in Relativity Theory: Arguments for a Philosophy of Being', in J. T. Frazer, *The Voices of Time*, George Braziller: New York, 1966.

26. P. L. Csonka, 'Advanced Effects in Particle Physics, I', *Physics Review* 180, no. 5 (1969).

27. James T. Culbertson, *The Minds of Robots*, University of Illinois Press: Urbana, 1963.

28. Alexandra David-Neel, *Magic and Mystery in Tibet*, Penguin: Baltimore, 1971.

29. Stanley R. Dean, 'Metapsychiatry: The Confluence of Psychiatry and Mysticism', *Fields Within Fields*, no. 11 (Spring 1974): 3–11.

30. B. S. DeWitt, 'Quantum Mechanics and Reality', *Physics Today* 23, no. 9 (1970): 30.

31. ———, 'Quantum-Mechanics Debate', *Physics Today*, April 1971.

32. Charles Eliot, *Japanese Buddhism*, Barnes & Noble: New York, 1969.

33. W. Y. Evans-Wentz, *The Tibetan Book of the Great Liberation*, Oxford University Press: New York, 1954.

34. Henri Eyraud, 'The Problem of the Infinite: Transfinite Numbers and Alephs', in F. Le Lionnais, *Great Currents of Mathematical Thought*, Dover Publications: New York, 1971.

35. David Finkelstein, 'The Space-Time Code, *Physical Review*, 5D, no. 12 (15 June, 1972): 2922.
36. Keith Floyd, 'Of Time and the Mind,' *Fields Within Fields*, no. 10 (Winter 1973–1974): 47–57.
37. Heinz Von Foerster, 'On Constructing a Reality', in *Environmental Design Research*, F. E. Preiser (ed.), vol. 2, Dowden, Hutchinson & Ross: Stroudsburg, Pa., 1973.
38. Vincent H. Gaddis, *Mysterious Fires and Lights*, Dell: New York, 1967.
39. Adolf Grünbaum, *Philosophical Problems of Space and Time*, D. Reidel: Boston, 1973.
40. Patrick A. Heelan, *Quantum Mechanics and Objectivity*, Martinus Nijhoff: The Hague, 1965.
41. Werner Heisenberg, *Physics and Philosophy*, Harper Torchbooks: New York, 1958.
42. Aniela Jaffé, *The Myth of Meaning*, Hodder and Stoughton: London, 1970.
43. J. M. Jauch, *Are Quanta Real?*, Indiana University Press: Bloomington, 1975.
44. Sir James Jeans, *The Mysterious Universe*, E. P. Dutton, New York, 1932.
45. ——, *Physics and Philosophy*, University of Michigan Press: Ann Arbor Paperbacks, 1958.
46. Carl G. Jung, *Flying Saucers*, Signet: New York, 1969.
47. ——, *Man and His Symbols*, Doubleday: New York, 1964.
48. Carl Jung and Wolfgang Pauli, *The Interpretation of Nature and the Psyche*, Bollingen Series LI, Pantheon Books: New York, 1955.
49. Arthur Koestler, *The Ghost in the Machine*, Henry Regnery: New York, 1967.
50. Gopi Krishna, *The Biological Basis of Religion and Genius*, New York, 1972.
51. ——, *Kundalina: The Evolutionary Energy in Man*, Berkeley, 1971.
52. Susan K. Langer, *Philosophy in a New Key*, Harvard University Press: Cambridge, 1942.
53. Lawrence LeShan, *The Medium, The Mystic, and the Physicist*, Viking Press: New York, 1974.
54. John C. Lilly, *The Human Biocomputer*, Bantam Books: New York, 1972.

55. Magoroh Maruyama and Arthur Harkins, *Cultures Beyond the Earth*, Vintage Books: New York, 1975.

56. Robert A. Monroe, *Journeys out of the Body*, Anchor Books/ Doubleday: New York, 1971.

57. Charles Muses and Arthur M. Young, *Consciousness and Reality*, Outerbridge & Lazard: New York, 1972.

58. John G. Neihardt, *Black Elk Speaks*, Pocket Books: New York, 1972.

59. André Padoux, *Recherches sur la Symbolique et l'Énergie de la Parole dans Certains Textes Tantriques*, E. de. Médicis: Paris, 1963.

60. Swami Panchadasi, *The Astral World*, 1921.

61. Joseph Chilton Pearce, *The Crack in the Cosmic Egg*, Pocket Books: New York, 1973.

62. ——, *Exploring the Crack in the Cosmic Egg*, Julian Press: New York, 1974.

63. Jean Piaget, *The Child and Reality*, Grossman: New York, 1972.

64. S. Pratyagatmananda, *The Metaphysics of Physics*, Ganesh: Madras, India, 1964.

65. Karl H. Pribram, *Languages of the Brain*, Prentice-Hall: Englewood Cliffs, N.J., 1971.

66. Carl Sagan, *The Cosmic Connection*, Dell: New York, 1973.

67. Jack Sarfatti, 'Implications of Meta-Physics for Psychoenergetic Systems', in *Psychoenergetic Systems*, vol. 1, Gordon and Breach: London, 1974.

68. Jack Sarfatti and Bob Toben, *Space-Time and Beyond*, E. P. Dutton: New York, 1975.

69. Satprem, *Sri Aurobindo or the Adventure of Consciousness*, Harper & Row: New York, 1968.

70. As quoted in Edwin Schlossberg, *Einstein and Beckett*, Links Books: New York, 1973.

71. J. R. Smythies, *Analysis of Perception*, Humanities: New York, 1956.

72. D. T. Suzuki, *On Indian Mahayana Buddhism*, Harper & Row: New York, 1968.

73. Michael Talbot, *A Mile to Midsummer*, work in progress.

74. Paul Twitchell, *The Tiger's Fangs*, Lancer Books: New York, 1969.

75. *Chuang Tzu*, trans. James Legge, arranged by Clae Waltham, Ace Books: New York, 1971.

76. P. J. Van Heerden, *The Foundation of Empirical Knowledge*, N. V. Uitgeverij Wistik: Wassenaar, The Netherlands, 1968.

77. Evan Harris Walker, 'The Nature of Consciousness', *Mathematical Biosciences* 7 (1970): 138–197.

78. Hermann Weyl, *Space-Time-Matter*, Methuen: London, 1922.

79. John A. Wheeler, 'Superspace and the Nature of Quantum Geometrodynamics', in C. DeWitt and J. A. Wheeler, *Battelle Rencontres, 1967 Lectures in Mathematics and Physics*, W. A. Benjamin: New York, 1968.

80. John A. Wheeler with C. Misner and K. S. Thorne, *Gravitation*, Freeman: San Francisco, 1973.

81. Alfred North Whitehead, *The Concept of Nature*, Macmillan: New York, 1925.

82. ——, *Science and the Modern World*, Free Press: New York, 1967.

83. Sir Edmund Whittaker, *Space and Spirit*, Regnery: Hinsdale, Ill., 1948.

84. Norbert Wiener, *God and Golem, Inc.*, M.I.T. Press: Cambridge, 1964.

85. ——, *The Human Use of Human Beings*, Avon Books: New York, 1967.

86. E. P. Wigner, *Symmetries and Reflections*, Indiana University Press: Bloomington, 1967.

87. John Wilson, 'Film Literacy in Africa', *Canadian Communications*, vol. 1, no. 4 (Summer 1961).

88. Sir John Woodroffe, *Mahamaya: The World as Power, Power as Consciousness*, Ganesh & Co.: Madras, India, 1964.

89. ——, *The Serpent Power*, Dover: New York, 1974.

90. J. Zimmerman, 'Time and Quantum Theory', in J. T. Fraser, *The Voices of Time*, George Braziller: New York, 1966.

NOTES
TO *1992* AFTERWORD

1. John Gliedman, 'Turning Einstein Upside Down', *Science Digest*, 92, no. 10 (October 1984), p. 96.

2. Fritz Rohrlich, 'Facing Quantum Mechanical Reality', *Science*, 221, no. 4, 617 (23 September, 1983), pp. 1, 251.

3. Basil Hiley, 'Quantum Mechanics Passes the Test', *New Scientist* (6 January, 1983), p. 19.
4. John Polkinghorne, *The Quantum World* (Harlow, Essex, England: Longman, 1984).
5. Malcolm W. Browne, 'Quantum Theory: Disturbing Questions Remain Unresolved', *New York Times* (11 February, 1986), p. C3.
6. Ibid.
7. P. C. W. Davies and J. R. Brown, *The Ghost in the Atom* (Cambridge, England: Cambridge University Press, 1986), p. 129.
8. John Gleidman, 'Interview with Brian Josephson', *Omni*, 4, no. 10 (July 1982), p. 88.
9. James Gleick, 'In Defiance of Einstein, Physicists Seek Faster-Than-Light Messages', *New York Times* (27 October, 1987), p. C3.
10. David H. Freedman, 'Weird Science', *Discover*, 11, no. 11 (November 1990), pp. 66–7.
11. Rupert Sheldrake, *The Presence of the Past* (New York: Times Books, 1988), pp. 304–307.
12. Stanislav Grof, *Beyond the Brain* (Albany, N.Y.: State University of New York Press, 1985).
13. F. David Peat, *Synchronicity: The Bridge Between Matter and Mind* (New York: Bantam, 1987).
14. Kenneth Ring, *Life at Death* (New York: Quill, 1980).
15. F. David Peat, *Einstein's Moon* (Chicago: Contemporary Books, 1990), p. 140.
16. Dietrick E. Thomsen, 'In the Beginning Was Quantum Mechanics', *Science News*, 131, no. 22 (30 May, 1987), p. 346.
17. Ibid, pp. 346–7; see also Sten Odenwalk, 'Einstein's Fudge Factor', *Sky & Telescope* (April 1991); and Ian Redmount, 'Wormholes, Time Travel, and Quantum Gravity', *New Scientist*, 126, no. 1714 (28 April, 1990).
18. Nick Herbert, *Faster Than Light* (New York: New American Library, 1988), p. 136.
19. 'Time Travel, Quantum-Style', *Science News*, 138, no. 10 (8 September, 1990), p. 159.
20. Malcom W. Browne, 'Three Scientists Say Travel in Time Isn't So Far Out', *New York Times* (22 November, 1988), p. C1 C7.

21. David J. Bohm, 'A New Theory of the Relationship of Mind and Matter', *Journal of the American Society for Psychical Research*, 80, no. 2 (April 1986).

22. Robert G. Jahn and Brenda J. Dunne, *The Margins of Reality: The Role of Consciousness in the Physical World* (New York: Harcourt Brace Jovanovich, 1987), p. 144.

23. Private communication with the author, 16 December, 1988.

24. Dietrick E. Thomsen, 'A Midrash Upon Quantum Mechanics', *Science News*, 132, no. 2 (11 July, 1987), p. 27.

25. Ibid.

26. Willis Harman, *Global Mind Change* (Sausalito, California: Institute of Noetic Sciences), p. 11.

27. George Wald, 'Cosmology of Life and Mind', *Los Alamos Science*, 16 (1988), pp. 10–11.

28. Grof, *op. cit.*, p. 89.

29. Morris Berman, *The Reenchantment of the World* (New York: Bantam Books, 1984), p. 141.

30. Larry Dossey, M.D., *Recovering the Soul* (New York: Bantam, 1989), pp. 2–8.

31. Harman, *op. cit.*, p. 34.

For Further Reading

The reader may also be interested in the books listed alphabetically by author in the Notes, above.

Arias-Larreta, Abraham, *Pre-Columbian Masterpieces*, Indo-American Library: Kansas City, Missouri, 1967.

Bancroft, Hubert Howe, *The Native Races*, History Company: San Francisco, 1886.

Barnard, G. C., *Samuel Beckett: A New Approach*, Dodd, Mead: New York, 1970.

Barnett, Lincoln, *The Universe and Dr. Einstein*, Bantam Books: New York, 1968.

Beals, Carleton, *Stories Told by the Aztecs*, Abelard-Schuman: New York, 1970.

Becker, Ernest, *Angel in Armor*, George Braziller: New York, 1969.

——, *The Birth and Death of Meaning*, Free Press of Glencoe: New York, 1962.

——, *The Denial of Death*, Macmillan: New York, 1973.

Belinfante, F. J., *A Survey of Hidden Variable Theories*, Pergamon Press: New York, 1973.

Bergman, P. G., *The Riddle of Gravitation*, Charles Scribner's Sons: New York, 1968.

Berner, Jeff, *The Innerspace Project*, World Publishing: New York, 1972.

Bertalanffy, Ludwig von, *Robots, Men and Minds*, George Braziller: New York, 1967.

——, 'The Theory of Open Systems in Physics and Biology', *Science*, 111 (1950): 23–29.

Besant, Annie, *The Ancient Wisdom*, Theosophical Publishing House: Adyar, India, 1897.

Bohm, David, *Quantum Theory*, Prentice-Hall: Englewood Cliffs, N.J., 1951.

——, *The Special Theory of Relativity*, Benjamin: New York, 1965.

Bondi, H. *Assumption and Myth in Physical Theory*, Cambridge University Press: New York, 1967.

Burland, C. A., *The Gods of Mexico*, Eyre & Spottiswoode: London, 1967.

Carrington, Hereward, and Muldoon, Sylvan J., *The Projection of the Astral Body*, Samuel Weiser: New York, 1970.

Colodny, Robert G., *Paradigms & Paradoxes*, University of Pittsburgh Press: Pittsburgh, 1972.

Coomaraswamy, Ananda, *The Dance of Shiva*, Straus & Cudahy: New York, 1957.

Crookall, Robert, *The Mechanisms of Astral Projection*, Darshana International: Moradabad, India, 1968.

Debergh, Joseph, and Sharkey, Don, *Our Lady of Beauraing*, Hanover House: New York, 1958.

DeWitt, Bryce S., and Graham, Neill, *The Many-Worlds Interpretation of Quantum Mechanics*, Princeton University Press: Princeton, N.J., 1973.

Duran, Fray Diego, *The Aztecs*, Orion Press: New York, 1964.

Einstein, Albert, *Essays in Science*, Philosophical Library: New York, 1934.

——, *The Evolution of Physics*, Simon and Schuster: New York, 1967.

——, *Ideas and Opinions*, Bonanza: New York, 1954.

——, *The Meaning of Relativity*, Princeton University Press: Princeton, N.J., 1970.

——, *Out of My Later Years*, Littlefield, Adams: Totowa, N.J., 1967.

——, *The World as I see It*, Philosophical Library: New York, 1949.

Eliade, Mircea, *Myth and Reality*, Harper & Row: New York, 1963.

Feynman, Richard, *The Character of Physical Law*, M.I.T. Press: Cambridge, 1965.

Gardner, Martin, ed., *Relativity for the Millions*, Macmillan Company: New York, 1962.

——, *Rudolf Carnap: Philosophical Foundations of Physics*, Basic Books: New York, 1966.

Gillett, H. M., *Famous Shrines of Our Lady*, Newman Press: Westminster, Md., 1952.

Heisenberg, W., *Physics and Beyond*, Harper & Row: New York, 1971.

Helle, Jean, *Miracles*, David McKay: New York, 1952.

Honoré, Pierre, *In Quest of the White God*, Lehrburger, Hutchinson: London, 1963.

Hume, Robert Ernest, *The Thirteen Principal Upanishads*, Oxford University Press: London, 1949.

Huxley, Aldous, *The Doors of Perception*, Chatto & Windus: London, 1960.

Irwin, Constance, *Fair Gods and Stone Faces*, St Martin's Press: New York, 1963.

Jaffé, Aniela, *Apparitions and Precognition*, University Books: New York, 1963.

Jammer, Max, *Concepts of Force*, Harper & Row: New York, 1962.

Jung, Carl G., *The Collected Works of C. G. Jung*, Bollingen series XX, Pantheon Books: New York, 1959.

Kubler, George, *The Shape of Time*, Yale University Press: New Haven, 1962.

Kuznetsov, Boris, *Einstein*, Phaedra: New York, 1970.

Le Lionnais, F., *Great Currents of Mathematical Thought*, Dover: New York, 1971.

Leonard, George B., *The Transformation*, Delacorte Press: New York, 1972.

Lilly, John C., *The Center of the Cyclone*, Bantam Books: New York, 1972.

Mackenzie, Donald A., *Myths of Pre-Columbian America*, Gresham: London, 1924.

Maeterlinck, Maurice, *The Great Secret*, University Books: New York, 1969.

McLuhan, Marshall, *The Gutenberg Galaxy*, Signet: New York, 1962.

Muller, Herbert J., *Science and Criticism*, Yale University Press, New Haven, 1943.

Munitz, Milton K., *Logic and Ontology*, New York University Press: New York, 1973.

O'Regan, Brendan, *Psychoenergetic Systems*, vol. 1, Gordon and Breach: London, 1974.

Progoff, Ira, *Jung's Psychology*, Doubleday Anchor: Garden City, N.Y., 1973.

Radhakrishnan, S., *The Principal Upanishads*, Harper & Row: New York, 1953.

Reichenbach, Hans, *The Theory of Relativity and a Priori Knowledge*, University of California: Los Angeles, 1965.

Roys, Ralph L., *The Book of Chilam Balam of Chumayel*, University of Oklahoma Press: Norman, 1967.

Shankaranarayanan, S., *The Ten Great Cosmic Powers*, Dipti Publications: Pondicherry, India, 1972.

Sinnett, A. P., *The Mahatma Letters*, Theosophical Publishing House: Adyar, India, 1972.

Stallo, J. B., *The Concepts and Theories of Modern Physics*, Harvard University Press: Cambridge, 1960.

Thompson, William Irwin, *At the Edge of History*, Harper & Row: New York, 1971.

——, *Passages About Earth*, Perennial Library: New York, 1974.

Van Fraasen, Bas C., *An Introduction to the Philosophy of Time and Space*, Random House: New York, 1970.

Woodroffe, Sir John, *Hymn to Kali Karpuradi-Stotra*, Ganesh: Madras, India, 1953.

——, *Principles of Tantra*, Ganesh: Madras, India, 1969.

Young, J. Z., *Doubt and Certainty in Science*, Oxford University Press: New York, 1960.

Yourgrau, Wolfgang & Mandelstam, Stanley, *Variational Principles in Dynamics and Quantum Theory*, W. B. Saunders: Philadelphia, 1968.

Zimmerman, J., 'The Macroscopic Nature of Space-Time', *American Journal of Physics*, v. 30, no. 2 (1962).

Index

PENGUIN

ARKANA

NEW AGE BOOKS FOR MIND, BODY & SPIRIT

With over 200 titles currently in print, Arkana is the leading name in quality books for mind, body and spirit. Arkana encompasses the spirituality of both East and West, ancient and new. A vast range of interests is covered, including Psychology and Transformation, Health, Science and Mysticism, Women's Spirituality, Zen, Western Traditions and Astrology.

If you would like a catalogue of Arkana books, please write to:

Sales Department – Arkana
Penguin Books USA Inc.
375 Hudson Street
New York, NY 10014

Arkana Marketing Department
Penguin Books Ltd
27 Wrights Lane
London W8 5TZ

Available while stocks last.

PENGUIN

ARKANA

NEW AGE BOOKS FOR MIND, BODY & SPIRIT

A SELECTION OF TITLES

Head Off Stress: Beyond the Bottom Line
D. E. Harding

Learning to head off stress takes no time at all and is impossible to forget – all it requires is that we dare take a fresh look at ourselves. This infallible and revolutionary guide from the author of *On Having No Head* – whose work C. S. Lewis described as 'highest genius' – shows how.

The Participatory Mind Henryk Skolimowski

In a Grand Theory of participatory mind that builds on the insights of such thinkers as Teilhard de Chardin and Bergson as well as contemporaries Dobzhansky and Bateson, Skolimowski points to a new order, one brought about by a Western mind returning to, then reintegrating, the spiritual. This quest for fresh perspectives, as we approach the twenty-first century, has now become 'the hallmark of our times'.

The Magus of Strovolos: The Extraordinary World of a Spiritual Healer Kyriacos C. Markides

This vivid account introduces us to the rich and intricate world of Daskalos, the Magus of Strovolos – a true healer who draws upon a seemingly limitless mixture of esoteric teachings, psychology, reincarnation, demonology, cosmology and mysticism, from both East and West. 'This is a really marvellous book . . . one of the most extraordinary accounts of a "magical" personality since Ouspensky's account of Gurdjieff' – Colin Wilson

The Great Year Nicholas Campion

The Great Year raises important questions concerning the nature and function of political prophecy in late twentieth-century society, whether it be the ideological fringes of the New Age Movement, mainstream political ideology or the extremes of Stalinism and Fascism. Are we, as some contemporary writers think, coming to the End of History? Or is the belief in Millenarianism and the imminent dawning of a New Age nothing more than a collective delusion?

PENGUIN

ARKANA

NEW AGE BOOKS FOR MIND, BODY & SPIRIT

A SELECTION OF TITLES

The Revised Waite's Compendium of Natal Astrology
Alan Candlish

This completely revised edition retains the basic structure of Waite's classic work while making major improvements to accuracy and readability. With a new computer-generated Ephemeris, complete for the years 1900 to 2010, and a Table of Houses that now allows astrologers to choose between seven house systems, it provides all the information on houses, signs and planets the astrologer needs to draw up and interpret a full natal chart.

A Time to Heal Beata Bishop

The inspiring story of a woman's triumph over life-threatening disease – through an unorthodox therapy. When Beata Bishop's cancer spread into the lymphatic system, she rejected the options of surgery or 'wait-to-die' and travelled to the Gerson clinic in Mexico, for therapy based on optimum nutrition and thorough detoxification. Over a decade later, she is fit and well, enjoying life to the full.

Tao Te Ching The Richard Wilhelm Edition

Encompassing philosophical speculation and mystical reflection, the *Tao Te Ching* has been translated more often than any other book except the Bible, and more analysed than any other Chinese classic. Richard Wilhelm's acclaimed 1910 translation is here made available in English.

The Book of the Dead E. A. Wallis Budge

Intended to give the deceased immortality, the Ancient Egyptian *Book of the Dead* was a vital piece of 'luggage' on the soul's journey to the other world, providing for every need: victory over enemies, the procurement of friendship and – ultimately – entry into the kingdom of Osiris.

Astrology: A Key to Personality Jeff Mayo

Astrology: A Key to Personality is designed to help you find out who you *really* are. A book for beginners wanting simple instructions on how to interpret a chart, as well as for old hands seeking fresh perspectives, it offers a unique system of self-discovery.

NEW AGE BOOKS FOR MIND, BODY & SPIRIT

A SELECTION OF TITLES

Weavers of Wisdom: Women Mystics of the Twentieth Century
Anne Bancroft

Throughout history women have sought answers to eternal questions about existence and beyond – yet most gurus, philosophers and religious leaders have been men. Through exploring the teachings of fifteen women mystics – each with her own approach to what she calls 'the truth that goes beyond the ordinary' – Anne Bancroft gives a rare, cohesive and fascinating insight into the diversity of female approaches to mysticism.

Dynamics of the Unconscious: Seminars in Psychological Astrology II
Liz Greene and Howard Sasportas

The authors of *The Development of the Personality* team up again to show how the dynamics of depth psychology interact with your birth chart. They shed new light on the psychology and astrology of aggression and depression – the darker elements of the adult personality that we must confront if we are to grow to find the wisdom within.

The Myth of the Goddess   Anne Baring and Jules Cashford

'This generous and ambitious study . . . is packed with knowledge from a heaped board of sources ... But like the cross-dressed heroine of a chivalric romance, it really records, not the story of the past, but a quest for the buried Grail of the feminine, which will heal the self-inflicted wounds humanity continues to open and reopen' – *Independent on Sunday*

The Hidden Tradition in Europe   Yuri Stoyanov

Christianity has always defined itself through fierce opposition to powerful 'heresies'; yet it is only recently that we have begun to retrieve these remarkable, underground traditions, buried beneath the contempt of the Church. In this superb piece of scholarly detective work Yuri Stoyanov illuminates unsuspected religious and political undercurrents lying beneath the surface of official history. 'The author's knowledge of relevant original sources is remarkable; and he has distilled them into a convincing and very readable whole' – Sir Steven Runciman

PENGUIN

ARKANA

NEW AGE BOOKS FOR MIND, BODY & SPIRIT

A SELECTION OF TITLES

A Course in Miracles
The Course, Workbook for Students and Manual for Teachers

Hailed as 'one of the most remarkable systems of spiritual truth available today', *A Course in Miracles* is a self-study course designed to shift our perceptions, heal our minds and change our behaviour, teaching us to experience miracles – 'natural expressions of love' – rather than problems generated by fear in our lives.

Fire in the Heart Kyriacos C. Markides

A sequel to *The Magus of Strovolus* and *Homage to the Sun*, *Fire in the Heart* centres on Daskalos, the Cypriot healer and miracle-worker and his successor-designate Kostas. The author, who has witnessed much that is startling in his years with the two magi, believes humanity may today be on the verge of a revolution in consciousness 'more profound than the Renaissance and the Enlightenment combined'.

The Western Way Caitlín and John Matthews

The Native Tradition and *The Hermetic Tradition* are now published together in one volume. The perennial wisdom of the Western Way has woven its bright and beckoning thread through religion, folklore and magic, ever reminding us of our connection with the earth mysteries of our ancestors and the mysticism of the Gnostic traditions.

Shamanism: Archaic Techniques of Ecstasy Mircea Eliade

Throughout Siberia and Central Asia, religious life traditionally centres around the figure of the shaman: magician and medicine man, healer and miracle-doer, priest and poet. 'Has become the standard work on the subject and justifies its claim to be the first book to study the phenomenon over a wide field and in a properly religious context' – *The Times Literary Supplement*

PENGUIN

ARKANA

NEW AGE BOOKS FOR MIND, BODY & SPIRIT

A SELECTION OF TITLES

Neal's Yard Natural Remedies
Susan Curtis, Romy Fraser and Irene Kohler

Natural remedies for common ailments from the pioneering Neal's Yard Apothecary Shop. An invaluable resource for everyone wishing to take responsibility for their own health, enabling you to make your own choice from homeopathy, aromatherapy and herbalism.

The Garden of the Prophet Kahlil Gibran

First published posthumously in 1933, this is the book Gibran was working on in the years leading up to his untimely death: the long-lost companion to his world-famous *The Prophet*. Full of insights expressed in Gibran's unique style on the nature of wisdom, time, loneliness and God, it resonates with humanity and compassion on every page.

The Anatomy of Fate Z'ev ben Shimon Halevi

The Anatomy of Fate, a book about spiritual roots and reasons, is one of the wisest and most learned books on astrology you are likely to read. It is a study of the nature of existence all the way from the atom up to the absolute, and of man's place as an individual and social creature in the scheme of things in between.

Power of the Witch Laurie Cabot

In fascinating detail, Laurie Cabot describes the techniques and rituals involved in charging tools, brewing magical potions and casting vigorous, tantalizing spells. Intriguing and accessible, this taboo-shattering guide will educate and enlighten even the most sceptical reader in the ways of an ancient faith that has much to offer today's world.

PENGUIN

ARKANA

NEW AGE BOOKS FOR MIND, BODY & SPIRIT

A SELECTION OF TITLES

Light on Life   Hart deFouw and Robert Svoboda

Jyotish or Indian astrology is an ancient and complex method of exploring the nature of time and space and its effect upon the individual. Formerly a closed book to the West, the subject has now been clarified and explained by Hart deFouw and Robert Svoboda, two experts and long-term practitioners.

The Moment of Astrology   Geoffrey Cornelius

'This is an extraordinary book ... I believe that within the astrological tradition it is the most important since the great flowering of European astrology more than three hundred years ago ... Quietly but deeply subversive, this is a book for lovers of wisdom' – from the Foreword by Patrick Curry

Homage to the Sun: The Wisdom of the Magus of Strovolos
Kyriacos C. Markides

Homage to the Sun continues the adventure into the mysterious and extraordinary world of the spiritual teacher and healer Daskalos, the 'Magus of Strovolos'. The logical foundations of Daskalos's world of other dimensions are revealed to us – invisible masters, past-life memories and guardian angels, all explained by the Magus with great lucidity and scientific precision.

The Eagle's Gift   Carlos Castaneda

In the sixth book in his astounding journey into sorcery, Castaneda returns to Mexico. Entering once more a world of unknown terrors, hallucinatory visions and dazzling insights, he discovers that he is to replace the Yaqui Indian don Juan as leader of the apprentice sorcerers – and learns of the significance of the Eagle.

PENGUIN
ARKANA

NEW AGE BOOKS FOR MIND, BODY & SPIRIT

A SELECTION OF TITLES

Being Intimate: A Guide to Successful Relationships
John Amodeo and Kris Wentworth

This invaluable guide aims to enrich one of the most important – yet often most problematic – aspects of our lives: intimate relationships and friendships. 'A clear and practical guide to realization and communication of authentic feelings, and thus an excellent pathway towards lasting intimacy and love' – George Leonard

River's Way   Arnold Mindell

River's Way presents us with the theoretical framework behind Arnold Mindell's two earlier studies of the psychology of body experience, *Dreambody* and *Working with the Dreaming Body*. In bridging divisions between psychotherapy, medicine and physics, it provides both client and therapist with a flexible and integrative approach to the total personality.

The Act of Creation   Arthur Koestler

This second book in Koestler's classic trio of works on the human mind (which opened with *The Sleepwalkers* and concludes with *The Ghost in the Machine*) advances the theory that all creative activities – the conscious and unconscious processes underlying artistic originality, scientific discovery and comic inspiration – share a basic pattern, which Koestler expounds and explores with all his usual clarity and brilliance.

Secrets of the Soil   Peter Tompkins and Christopher Bird

In this long-awaited sequel to their bestselling *The Secret Life of Plants* Peter Tompkins and Christopher Bird explore the revolutionary methods of biodynamic agriculture introduced by the scientist–philosopher–mystic Rudolf Steiner. They show how Steiner's astonishing 'homeopathic' fertilizers and growing techniques have been used to revitalize previously barren areas and to achieve amazing feats of productivity.

PENGUIN
ARKANA

NEW AGE BOOKS FOR MIND, BODY & SPIRIT

A SELECTION OF TITLES

The Dreambody in Relationships Arnold Mindell

All of us communicate on several levels at once, and Mindell shows how much of our silent language conflicts with overt behaviour. He argues that bringing all the hidden parts of ourselves to awareness as they affect us is important for the well-being not only of our relationships but also of the community – indeed, the world – in which we live.

The Sacred Yew   Anand Chetan and Diana Brueton

Recently it has been discovered that the yew can live for many thousands of years. *The Sacred Yew* is the inspiring story of one man's crusade to preserve this revered yet threatened tree and explain its importance to all our lives.

Be As You Are Sri Ramana Maharshi

'The ultimate truth is so simple.' This is the message of Sri Ramana Maharshi, one of India's most revered spiritual masters whose teachings, forty years after his death, are speaking to growing audiences worldwide. 'That sense of presence, of the direct communication of the truth so far as it can be put into words, is there on every page' – *Parabola*

In Search of the Miraculous: Fragments of an Unknown Teaching P. D. Ouspensky

Ouspensky's renowned, vivid and characteristically honest account of his work with Gurdjieff from 1915 to 1918. 'Undoubtedly a *tour de force*. To put entirely new and very complex cosmology and psychology into fewer than 400 pages, and to do this with a simplicity and vividness that makes the book accessible to any educated reader, is in itself something of an achievement' – *The Times Literary Supplement*

PENGUIN

ARKANA

NEW AGE BOOKS FOR MIND, BODY & SPIRIT

A SELECTION OF TITLES

On Having No Head: Zen and the Rediscovery of the Obvious
D. E. Harding

'Reason and imagination and all mental chatter died down ... I forgot my name, my humanness, my thingness, all that could be called me or mine. Past and future dropped away ...' Thus Douglas Harding describes his first experience of headlessness, or no self. This classic work truly conveys the experience that mystics of all ages have tried to put into words.

The Book of Chuang Tzu
Translated by Martin Palmer with Elizabeth Breuilly

The Book of Chuang Tzu draws together the stories, tales, jokes and anecdotes that have gathered around the figure of Chuang Tzu, entering into debate with logic, dancing around philosophy, making Confucian earnestness – along with the pretensions of emperors, bureaucrats and sages – the frequent butt of its jokes.

Money, Heart and Mind   William Bloom

This book finally delivers what many people have long wanted: an illuminating, intelligent and compassionate understanding of money. Sharply perceptive about our psychological needs, William Bloom also enlightens us about the true social context of money and encourages us to transform not only our personal but also our organizational and political financial behaviour.

The Universe Story   Brian Swimme and Thomas Berry

Here is the unique story of our human development – biological, social, intellectual and spiritual – alongside the evolution of all other species, the future of each resting now on resolution of the tension between humans exploiting dwindling resources for their own benefit and those dedicated to an 'ecozoic' age in which the well-being of the entire Earth community will be of primary concern.

PENGUIN

ARKANA

NEW AGE BOOKS FOR MIND, BODY & SPIRIT

A SELECTION OF TITLES

Daimonic Reality Patrick Harpur

Mysterious lights in the sky, phantom animals, visions of the Virgin Mary, UFOs, fairies, alien abductions ... Such anomalies have appeared throughout the centuries and, despite the denials of Church and Science, continue to be reported all over the world. 'A startling "field guide to the Otherworld" which should stop even the sceptical in their tracks' – *Observer*. 'A brave, thought-provoking book' – *Daily Mail*

The Second Ring of Power Carlos Castaneda

Carlos Castaneda's journey into the world of sorcery has captivated millions. In this fifth book, he introduces the reader to Dona Soledad, whose mission is to test Castaneda by a series of terrifying tricks. Thus Castaneda is initiated into experiences so intense, so profoundly disturbing, as to be an assault on reason and on every preconceived notion of life ...

Dialogues with Scientists and Sages: The Search for Unity
Renée Weber

In their own words, contemporary scientists and mystics – from the Dalai Lama to Stephen Hawking – share with us their richly diverse views on space, time, matter, energy, life, consciousness, creation and our place in the scheme of things. Through the immediacy of verbatim dialogue, we encounter scientists who endorse mysticism, and those who oppose it; mystics who dismiss science, and those who embrace it.

Women in Search of the Sacred Anne Bancroft

This fascinating book surveys the careers of ten very different women and examines the ways in which they have developed their spiritual lives. Some, for example writer Susan Howatch, find that serving God is the key to a spiritually fulfilling life, whereas for Danah Zohar, a convert to Judaism, the sacred mystery of existence is allied to quantum physics.